Molten Salt Techniques

Volume 2

Molten Salt Techniques

Volume 2

Edited by
Robert J. Gale
and
David G. Lovering
Louisiana State University
Baton Rouge, Louisiana

Plenum Press • New York and London

Library of Congress Cataloging in Publication Data

Main entry under title:

Molten salt techniques.

Includes bibliographical references and index.
1. Fused salts. I. Lovering, D. G. (David G.) II. Gale, Robert J., 1942–
QD189.M59 1983 546′.343 83-9582
ISBN 0-306-41549-6 (v. 2)

A Division of Plenum Publishing Corporation
233 Spring Street, New York, N.Y. 10013

Printed in the United States of America

TO
MARY AND MARILYN

Namque tu solebas
Meas esse aliquid putare nugas
–Catullus

Contributors

Harry B. Bell • Department of Metallurgy, University of Strathclyde, Glasgow, Scotland

Robert J. Gale • Department of Chemistry, Louisiana State University, Baton Rouge, Louisiana

Trevor R. Griffiths • Department of Inorganic and Structural Chemistry, The University, Leeds, United Kingdom

Douglas Inman • Department of Metallurgy and Materials Science, Imperial College, London, United Kingdom

Frédéric Lantelme • Laboratoire d' Electrochimie, Université Pierre et Marie Curie, Paris, France

David G. Lovering • Department of Chemistry, Louisiana State University, Baton Rouge, Louisiana

Hansraj C. Maru • Energy Research Corporation, Danbury, Connecticut

Krishnan Rajeshwar • Department of Chemistry, University of Texas at Arlington, Arlington, Texas

Iain Sommerville • Department of Metallurgy and Materials Science, University of Toronto, Toronto, Canada

Foreword

This second volume carries on the excellent work of its predecessor, extending its scope to other melts and to other techniques. It continues to present first-hand understanding and experience of this difficult and demanding field. There is ever present the trade-off or reconciliation between the novel chemistry of systems not dominated by the mediating influence of a supposedly indifferent solvent and the high temperatures required to effect the fluidity of the system. At the limit, the very high temperatures so increase the rates of all reactions as to dissolve the temporal difference between the thermodynamic and the kinetic view of chemistry. What can happen will happen and invariably does happen. Vessels corrode, the apparatus becomes a reactant, and the number of tolerant materials able to withstand the attack shrinks to graphite, boron carbide or, if all else fails, to frozen parts of the molten salt itself.

It is probably true that there is no limit to man's ingenuity but I believe that God gave us molten salts just to test that thesis. If there is ever a Molten Salt Club, and Englishmen love clubs, its membership will be exclusive. It would certainly include the authors of this series.

University of Strathclyde Graham Hills

Preface

In the first volume of this series, we expressed our contention that a real need existed for practical guidance in the field of molten salt experimentation. We had envisaged editing only two volumes initially, then later wondered whether a further issue might be justified. By the time Volume 1 had gone to press, it had become evident that neither two nor three volumes would serve the requirements and enthusiasm of our colleagues. Indeed, we have been sufficiently encouraged by the responses of potential readers as well as contributors that the contents of at least six volumes have already been sketched out. Perhaps a few will throw up their hands in horror when they read of such an explosive expansion of the literature. However, we feel most strongly that the dissemination of practical advice in such an experimentally difficult area of study is fully justified. We prefer an explosion in print to an explosion in the laboratory or plant—and the latter events *are* still happening all too frequently. This series is not just another review, a theoretical concept, but an attempt to provide hard, factual guidance of a practical nature.

This volume continues the pattern set by the first issue; both melt types and techniques are featured. There are fewer chapters in the present volume solely because two major techniques—spectroscopy and electrochemistry—needed lengthier than average coverage. In future issues we propose to include shorter than average chapters, where appropriate, in order that nonroutine approaches and modifications for measuring physical properties, e.g., density, viscosity, refractive index, etc. may be described. We would welcome offers of contributions of this type.

Industrial interest in molten salts seems to have waned a little in recent years with the depressed world economic situation. This effect has been superimposed upon an inherent inertia to change which arises from sound commercial reasons associated with capital investment in conventional technology and accumulated expertise therein, as well as from a human factor. Nevertheless, expansion into new areas of molten salt studies at a fundamental and academic level certainly continues apace. It is important for this to happen in order that a storehouse of knowledge can provide for future progress in advanced technologies. An ability to experiment with new techniques and new systems must be considered a vital component in our armory.

In this and future volumes we intend to re-emphasize safety matters. We do not believe that hazards encountered with molten salts are any greater than in any other area—just less familiar. Vigilance is the by-word, caution the maxim, and factual knowledge the goal. If this series goes some way towards preventing unnecessary accidents amongst our friends and colleagues, the effort will have been worthwhile.

The bulk of our editorial and contributions to the present text were carried out in the Chemistry Department at Louisiana State University; one of us (D.G.L) completed an overseas attachment at that institution during the production stages. We would again offer our sincere appreciation to Chairman Daly (L.S.U), fellow faculty and staff for their interest, cooperation, and encouragement. The New York staff of Plenum, as well as their London counterparts, have again been helpful in the venture, particularly Jeanne Libby, Ellis Rosenberg, and Ken Derham. A special word of thanks must also go to all Louisiana friends and family in suffering our preoccupations during this period.

Robert J. Gale
David G. Lovering

Contents

Chapter 5. Electrochemistry - I

Frédéric Lantelme, Douglas Inman, and David G. Lovering

1

Introduction

Robert J. Gale and David G. Lovering

1. Research in Molten Salts

Any experimental chemist contemplating using molten salts—whether small-scale laboratory investigations or large-scale industrial operations are envisaged—will rapidly become aware of the diverse new practical skills to be acquired. Novel engineering designs, elaborate glassblowing, extreme materials specifications, as well as finely developed appreciation of basic chemical principles (including corrosion problems) will force newcomers to the field to adopt a reoriented outlook. Thus, a multitude of unfamiliar difficulties, that may require unique solutions, must be overcome when building a new molten salt facility. At least three areas of separate expertise need to be confronted:

- The Molten System **per se** — its chemistry and that of all materials, solutes associated with it.
- The Hardware — equipment to run at elevated temperatures in aggressive environments, possibly at reduced or increased pressure.
- The Measurements — application and control of signals, data collection and processing, preferably with automation.

Traditionally, molten salt chemists have tended to work either with a single anion or group of anion systems, e.g., halides, nitrates, etc., and/or with a particular project application,[1] e.g., nuclear fuel processing, metals extraction, etc. For example, halide eutectics of the alkali metals find application in metals extraction, batteries, and nuclear fuel reprocessing, while nitrate eutectics of the same metals have been used in metals treatment, organic nitrations, and thermal energy

Robert J. Gale and David G. Lovering • Chemistry Department, Louisiana State University, Baton Rouge, Louisiana 70803, U.S.A.

storage.[2] Consequently, in the laboratory the necessary handling and evaluation expertise is developed for a restricted field of study or application and each molten salt may be considered as an individual system requiring unique considerations. Sufficient experimental complexities arise for each system as to merit individual treatments being developed for them.

As a result of the practical limitations for molten salt studies within any one laboratory, a concentration of expertise for certain systems is apparent. It is, therefore, beneficial, even for skilled practitioners, to be exposed to methodologies developed for alternative systems and to learn of ancillary practical experiences. As techniques and technologies advance, developments in one area can be incorporated with another. A general awareness in the field of molten salt studies is clearly to be commended. Seemingly minor effects such as side-reactions and trace corrosion, or the presence of impurities, can exert major influence on the course of reactions in molten salts. A number of well-documented misinterpretations of experimental results have recently been uncovered in the literature; the catalytic role of interfering species such as water, oxides, and container corrosion products had not previously been recognized. Although borosilicate glassware may be used to ~450°C, for instance, it is necessary to be aware that this material can exhibit ion exchange, catalysis effects, and even impurity occlusion. Not being an entirely homogeneous substance, it will contain traces of dissolved air, water, and free oxides. At elevated temperatures, these substances can migrate relatively rapidly toward the melt interface. At that point they may partition or react with the melt; impurity metal ions can undergo leaching into the melt. Melts such as fluorides, hydroxides, carbonates, and others will react in bulk with glasses and with many ceramics, especially in the presence of air and water vapor. Unfortunately, pressures to analyze and report bulk effects have all too often meant that possibly controlling side-reactions have been overlooked or gone unrecorded. In molten salt practice, the experimenter cannot be warned too strongly to be watchful of such matters and to cast aside philosophies and preconceptions based on experiences with other media. Indeed, such attention to detail may indicate a program of trial and error, of materials assessment, and of apparatus development entailing substantial cost and time. Much of the day-to-day information and advice is passed on verbally within research groups themselves; from the experienced to the less experienced. Information of this type is disseminated by visitors and exchange arrangements as well. Molten salt laboratory groups, therefore, acquire skills and techniques for working with particular chemical systems in certain temperature ranges, for handling very small or very large quantities of melts with or without special atmosphere control and, generally, using one set of physical and/or evaluation techniques more than others, e.g., mainly spectroscopic, electrochemical, etc.

Having outlined some of the considerations facing potential molten salt researchers, it is easy to see why any particular group tends to acquire those skills appropriate to one or two melt systems, operating within well-defined temperature limits and using just a few investigative techniques. Despite these restrictions, it is still necessary to exercise a wide range of practical, engineering, and chemical knowledge. The choice and construction of furnaces with suitable temperature controllers, calibration, fail-safe and electrical/electronic safety, screening and power supply can take up to six months. The building of vacuum lines, dry boxes, and preparative facilities can, likewise, commit precious man-hours and require glassblowing, pipe-fitting, and engineering. Problems and costs are multiplied if corrosive gases (e.g., F_2, Cl_2), corrosive melts (e.g., F^-, CO_3^{2-}), or noxious solvents have to be employed, since stainless steel, perhaps Teflon coated, may be called for in construction. Anyone who has equipped a laboratory to work in such manner can attest that a large number of peripheral and support activities must be attended to and be in good working order, before the fundamental task of making measurements can commence. So, frequently, the molten salt chemist must modify commercial instruments to suit his or her specific needs. For example, one cannot simply take an nmr tube, fill it with a molten salt solution, and proceed to the spectrometer to record data, if a high temperature is required to maintain the system in a liquid state. It is likely to take several months, or more, to modify the equipment in some instances. In others, the measurements have to be abandoned because the technique essentially is unavailable. A few other examples, chosen at random, illustrate the problems to be faced:

- optics need reversing for i.r. spectrometers, due to black-body radiation interferences from hot melts.
- permanent magnets need extensive protective cooling for nmr or susceptibility measurements, if degaussing is to be prevented; alternatively, expensive electromagnets with attendant control problems have to be adopted.
- noninductively wound resistance furnaces are particularly difficult to wind if the zone of uniform temperature is very long; often earthed screens are essential.

Some of these topics are considered in chapters in Volumes I and II of *Molten Salt Techniques;* others will be dealt with in succeeding volumes. Some aspects of equipment specification and construction may be common to other areas of chemical research and advice is available elsewhere. For example, the ability to handle, modify, and develop modern electronic instruments and data handling hardware is a desirable prerequisite, but not unique to this field. Progress in molten salt research requires patient application and, above all, painstaking attention to the detail of hardware and support equipment. The next section is intended to assist the neophyte to get started. Although peer guidance is the optimum form of advice, several excellent texts of laboratory technique can be consulted before final practical decisions are made and the creative task of preparing the experiment commenced.

2. Establishing a Molten Salt Facility

2.1. Safety in the Laboratory

The first consideration, of course, is the chemical system and what has to be achieved or evaluated experimentally. We shall assume that a thorough literature search and study has resulted in the conclusion that a molten salt system has to be employed. This stage can involve consultation of general inorganic and organic texts, e.g.,[3,4] a search of chemical abstracts, and a review of the pertinent molten salt literature (e.g., (1), Chapter 1, reference list). If suitably specific profiles can be defined, computer literature searches may be helpful, though they are rarely comprehensive. An important early consideration is the reactivity and potential hazards associated with chemicals to be used. It is worth reiterating that chemicals in any form can be safely used if their properties are known and the necessary precautions observed.[2] There can be no substitutes for knowledge, understanding, and caution. In molten salt chemistry sometimes rather less knowledge than considered desirable may be available and, additionally, larger than usual quantities of solvents must be handled under more extreme conditions than required for "conventional" experimentation. We make no apologies for developing the safety theme further in this second volume. A rather lively debate has followed the report of an explosion of a nitrite/thiocyanate molten salt bath.[5] The fused salt bath of about 4 lb of materials, apparently being used as a heat-transfer medium, detonated with the estimated force of 1 lb dynamite causing extensive damage to a laboratory (Figure 1). The majority of general molten salt texts and practitioners **do emphasize** that potential dangers must be researched and, for example, Lovering has commented "that nearly all of these contingencies are known, well-characterized, and have technological solutions".[2] Apparently, literature comments on the danger of heating alkali nitrite–thiocyanate mixtures are not difficult to find either.[6] A variety of reference manuals and handbooks are helpful in researching safety aspects and assessing hazards (e.g., 1,7–10). To these we add a few common sense guidelines:

- Never mix, grind, or heat any salts without due consideration of their stability and reactivity. Some materials react together in the solid state, others explode, etc.
- Avoid contacting oxidizing agents with reducing agents (including organic solvents, metals, etc.), unless their interactions are documented and well-characterized. Due note needs to be taken of oxidizing or reducing impurities that may be present.

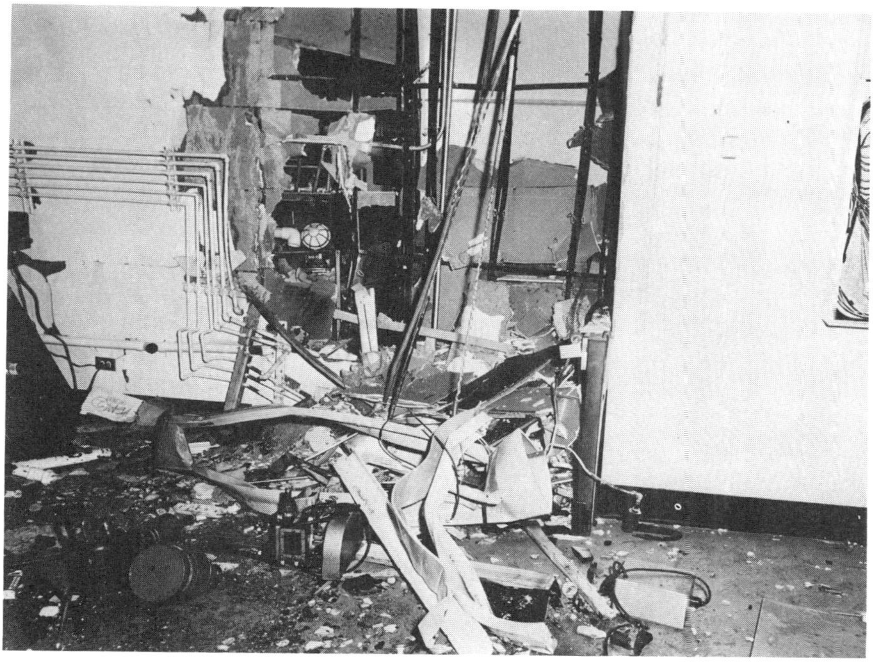

FIGURE 1. The scene of a Berkeley lab after a nitrite/thiocyanate melt explosion.

- Anticipate accidental mixtures arising. Secondary containers or shields can prevent incompatible materials contacting from breakages, e.g., glass vessels in oil baths, etc.
- If in doubt, or without information, experiment on a small scale with adequate safeguards (shielding, protective clothing, correct fire-fighting procedures).
- Never leave potentially dangerous systems unattended and inadequately posted.
- Avoid working alone because an added danger of fire and/or electrical hazards exists with high-temperature equipment.
- Be wise before an event! READ, CONSULT, THINK!

In the event of fires, provided that no danger is involved, the fire should be extinguished with fire-fighting equipment. If alone, or if there is any doubt, immediately seek assistance or activate the emergency alarm system. Molten salt chemicals which require special extinguishers should be identified prior to experimentation and the necessary equipment be placed at hand. Apparatus and laboratories should be adequately labelled and the name, address, and telephone number of the supervisor be displayed prominently. Fires may present special problems when working with molten salts because some conventional fire-fighting aids aggravate, rather than reduce, the dangerous situation. The necessary equipment for dealing with emergencies should have been located in a convenient (safe) position and identified *before* the first experiment is commenced. It may even be necessary in certain large-scale operations to deactivate conventional sprinkler systems to prevent their involvement. Where conventional and molten salt fire-fighting may be required, the situation can be complex. Table I contains the usual range of fire-fighting aids and some general advice. However, there is no universal rule—each system should be considered separately on its own merits. Since water is so commonly used to extinguish fires, it should be noted that its use for molten salt situations is not recommended! Explosive boiling and the splattering of hot melt invariably results in a worse situation to control. Forethought should be given if the fire (or potential fire) will involve electrical apparatus for heating or measurements and the danger of electric shock. It may be necessary to turn off the power suddenly and it is as well to be very familiar with the laboratory circuitry and external mains on/off switchbox. Master switches are the best grouped and conveniently located. Spilt melt often is easily absorbed using the materials indicated in Table I and, in some cases, it is expedient to allow fires to burn themselves out.

An obvious hazard is chemical burns from spilled or splashed molten liquids. If a very small area is involved, usually flooding the area of skin affected with large quantities of water is the treatment of choice, for 10 minutes or longer. On the other hand, in certain situations it may be necessary to resist the temptation to treat the splash with water and, firstly, remove the surplus melt with an absorbent. Flooding the area with water may enhance explosive boiling, steam, and even the release of toxic fumes, e.g., from cyanides. Strongly oxidizing melts cannot be permitted prolonged contact with skin and most burn dressings are organic in nature, like skin and tissue itself! First aid and burn kits should be close by and, however serious the burn, the correct treatment is no more than covering the wound with loosely applied dry sterile dressings. All burns other than of a trivial nature should be referred to professional medical attention.

Having made these points, it is to be hoped the operator will be even more aware of the need for prevention. Spills and burns are avoided by the use of proper handling procedures. A properly equipped molten salt laboratory should contain asbestos (or mineral substitute) gloves, slabs, trays, etc., together with tongs, clamps, platforms, and so on, with which to handle and place hot apparatus. It is usual to let a melt cool and solidify before attempting to dispose of it but it may require ladling into empty tins or suitable disposable containers, to prevent damage to glass cells upon expansive crystallization. Remote handling facilities are recommended wherever possible.

TABLE I.

A. Some hazards arising in molten salt experimentation

Hazard	*Associated problems*
Fires	Aggravated spillage Electrical insulation breakdown Flooding Toxic fumes Burns (chemical and physical)
Explosions	Dispersion of liquid melt Rupture of melt vessels Fire Facial and other bodily injury
Fumes	Toxicity Corrosion
Chemicals	Release of strong oxidizing or reducing agents Reaction with ancillary construction or support materials Personal acute danger

B. Recommended materials for handling molten salt hazards

Fire or spillage	Asbestos (and mineral substitute) blanket, sheets, fibers, etc. Sand, ceramic powder, refractory matting, glass fibers Minerals (mica), etc.[a]
Chemical/fume/other	Individual consideration such as ventilation, masks and breathing apparatus, enclosed atmosphere work

C. Materials to avoid with molten salt hazard

- Water
- Organic or inorganic gases and foams
- BCF and similar powders
- Incompatible chemical agents

[a] Manufacturers and suppliers can assist the selection of suitable agents and protective gear.

2.2. *Equipment Aspects*

Once the broad chemical parameters associated with the proposed program have been delineated, attention may be turned to physical aspects concerning the experimental hardware. It is usual to decide upon a set of preliminary physical parameters for the experiment when researching the conditions desired to effect a chemical reaction. Or, it may be that one or several physical properties of a system have to be measured as a function of another physical constraint.

The temperature range of operation will generally be dictated by the melt systems and the nature of the investigations to be undertaken. Thus, the type of heating systems and applicable materials are broadly defined at the outset. As an historical aside, it is interesting to note that in an essay describing chemical laboratories in the seventeenth and eighteenth centuries, Eklund[11] has remarked that one of the primary requirements of the adept chemist at that time was that he know his heats, both high and low. Charcoal filled furnaces of the period produced high heats, woods gave lower ones, and temperatures could be controlled crudely by flues and by designing separate chambers at varying distances from the main heat source. Water baths **(balneo mariae)** and the sand bath were utilized too, the latter requiring alertness to maintain constant temperature. The bunsen burner (R.W. Bunsen, 1811–1899) was a major breakthrough as a clean, convenient heat source, and electric resistance heaters were introduced only shortly before 1900.[12,13] Marsh, in about 1910, is accredited with long life Ni–Cr alloy wires for electrical resistance furnaces. These still remain the most popular choice for self-built tube furnaces, as well as many commercial designs. Some aspects of their construction and operation are considered in detail in other chapters in this volume. Other forms of heating are employed in special circumstances and a list of some of the alternatives available is given in Table II. Furnace types and materials are discussed generally in references 14–16 together with temperature measurement and control methods. Systems used with particular anion and melt classes can be found in the pertinent sections of this series.

Many other routine and not-so-routine operations are required of the molten

TABLE II. Methods to Achieve Elevated Temperatures

Heat source	Type of equipment
Chemical fuel	Traditional flame applied directly or through heat exchange medium. Important industrially. Control difficulties.
Electricity	Joule heating via a resistant element. •Furnace tube, e.g., wire on quartz, etc. •Rod in metal block, e.g., Al, brass, cast-iron •Cable or rod in melt, e.g., Drawtemp, etc. •Air oven •Heat-transfer fluid, e.g., water, silicone, liquid paraffin, organic vapor RF induction heating (remote) Fluidized bed heaters
Infra-red/Solar	Industrial/domestic solar heat collectors. Solids, liquids, and phase-change materials.
Light	Lasers. Specialist probes and local heating.
Nuclear	Fission and fusion. Heat transfer media may include liquid metals or molten salts.

salt experimentalist. Vacuum line manipulations, pressure measurements, the use of inert atmosphere glove boxes, and many other extremely helpful practical laboratory hints and suggestions can be found in Shriver's monograph *The Manipulation of Air-Sensitive Compounds*.[17] More recent chapters can be found on laboratory heat transfer, pumps and flow measurement, glove box technique, and vacuum technique in the excellent series "Techniques of Chemistry," Volume XIII, edited by Weissberger, Rossiter, *et al.*[18] Some specialist monographs are available, too, to help with glassblowing.[19,20]

2.3 Experimental Procedures

Molten salt experimentation has been and continues to be remarkably diversified. In the earlier review of experimental techniques for fused salts by Bailey and Janz,[14] the major areas are broadly divided into four:

I. Electrochemical
II. Spectroscopic
III. Transport, and
IV. Thermodynamic

Far more areas of current interest are discernable at present, although clear-cut categories cannot always be delineated; for example, the newer spectroelectrochemical techniques in melts, pioneered by Mamantov and co-workers and described in Volume I of this series, overlap both the electrochemical and spectroscopic areas. Insofar as some studies are separable, Table III outlines a few relevent areas, with no attempt at comprehensiveness or precise categorization. The newer electrochemical, spectroscopic, and surface studies developed in aqueous solutions now are being adopted for molten salts as new materials and equipment permit. In some cases, adoption is readily accomplished, for example, if fluidity, optical transparency, or conductivity are determining parameters; in other cases, the elevated operating temperatures, corrosivity, and reactivity may render some experimental techniques impossible.

The following approach is recommended for all new molten salt experimentation. Firstly, refer to the molten salt literature to see if the problems associated with the particular technique have been resolved. Even if literature reports are available for experimental procedures, the fields of study are usually sufficiently novel and undocumented that new approaches are regularly described and earlier results subjected to new interpretations. Constant literature scans are strongly advised. One attraction of molten salt work is that newcomers can make significant contributions by way of both new investigations and reappraisals of earlier work. Finally, do not be afraid to experiment! There is always a first time for everything.

TABLE III. Techniques Applicable to Molten Salt Studies

A. Electrochemical	
Potentiometric	Reference electrodes, membranes.
Voltammetric	Classical polarography, cyclic voltammetry, pulse and differential pulse methods, chronoamperometry, rotating disk (ring disc), ac and impedance techniques, stripping.
Galvanostatic	Chronopotentiometry and coulometry.
Miscellaneous	Surface studies by capacitance and double layer methods, computerized applications of the above.
B. Spectroelectrochemical	
Absorption	UV, visible and ir regions with transmission and reflectance geometries. Esr and nmr methods.
Luminescence	Raman scattering (SERS), electrogenerated fluorescence and chemiluminescence.
X-ray diffraction	**in situ** techniques at electrodes not yet applied to molten salt systems.
C. Spectroscopic	
Electronic	Visible and UV spectra; absorption and emission.
Vibrational	Infrared and Raman, Fourier analyses.
Spin, X-ray and Neutron diffraction	Also particle and nuclear studies.
D. Transport	
Transference	New interest in theoretical aspects of complex containing melts.
Diffusion	Electrochemical and tracer methods.
Viscosity	Reviews available of main techniques used, e.g., Ref. 14.
E. Thermodynamic	
Phase equilibria	Liquidus properties and thermal analyses.
Calorimetry	Specialized high-temperature calorimeters.
Cryoscopy	Colligative property and freezing point studies.
Surface Tension	Both detachment and maximum bubble pressure techniques utilized.
Density	Review available (21).
Phase separation	Vapor phase chromatography and adsorption methods.
F. Miscellaneous	Many other techniques have been applied to study such properties as compressibility by ultrasonic velocities, refractive index, magnetic susceptibilities, thermal conductivities, etc.

TABLE III (continued)

G. Simulation	Computer simulation is being applied increasingly to predict structure, transport and perhaps interfacial properties in melts. However, all such approaches depend both on adequate models as well as fundamental **experimental** data, e.g., pair potentials.

3. Future Prospects

All of chemical research has benefitted immensely from the advances in electronic instrumentation and computer facilities; molten salt studies are no exception. Nevertheless, progress in the application of modern laboratory instrumentation to molten salts often does lag other fields because of the real, practical difficulties of conducting experiments. Materials constraints are a major problem. Additionally, regular maintenance of instruments is essential, since the corrosive fumes usually present in molten salt laboratories can, for example, attack the solder joints rendering equipment unreliable or inaccurate. In large-scale plants remote location of control equipment is possible. As these objections are overcome, molten salt experimental work will continue to provide rewarding data and knowledge. Many of the experimental discoveries in research laboratories, particularly those relating to materials, find value in industrial or large-scale operations. At present, there is considerable interest in the application of molten salts in the power generation and storage industries,[2] including batteries (for traction and load-leveling, as well as reserve), fuel cells, thermal/solar energy and transfer, nuclear power. New areas of metals and semiconductor processing, as well as established technology of this genre, show promise.[2] Some current areas of interest are collected in Table IV.

Prospective developments in materials analysis and processing include high-temperature chromatography, two phase partitioning, and new separative procedures. The availability of new dimensionally and chemically stable substrates, in and upon which molten salts can be dispersed, opens up new possibilities for their use as catalysts. When suitable materials and handling facilities can be engineered, new pyrolitic routes for hazardous substances disposal, as well as reprocessing, will emerge. Every study of molten salts involves a plethora of techniques, and with improved methodology both data interpretation and progress are easier. Such studies of novel systems and problems require both use of the older, well-established methods and the development of new ones. The viability of new processes rests on adequate advances in parallel technologies, especially materials; this includes the discovery of economically sound production routes. To some extent, we have been waiting at the crossroads for progress in other areas to catch up. This can be a slow and frustrating process. Hopefully, with

TABLE IV. Current Areas for Molten Salt Activity and Research

Field of application	Main techniques
Materials	
Metals and semiconductor processing	Electrochemical and spectroelectrochemical, general properties.
Fuel chemistry and waste treatment	Separations, spectroscopic, mechanism, and kinetic studies.
Glasses and fluxes	Diverse physical and chemical properties.
Power	
Molten salt batteries	Electrochemical mechanism and transport, thermodynamic.
Fuel cell evaluation	Miscellaneous surface and general properties, mechanistic studies.
Thermal energy storage	Thermal and physical properties, stability and degradation, corrosivity.
Nuclear	Heat-exchange, separations chemistry, and radiation chemistry methodology.

new understandings, modern techniques, and undoubted enthusiasm the progress in molten salt techniques and technology can continue.

References

1. D.G. Lovering and R.J. Gale, *Molten Salt Techniques*, Vol. I, New York, Plenum Publishing Corp. (1983).
2. D.G. Lovering (Ed.), *Molten Salt Technology*, Chapter 1, New York, Plenum Publishing Corp. (1982) p. 6.
3. C.A. Jacobson (Ed.), *Encyclopedia of Chemical Reactions*, Vols. I-VIII, New York, Reinhold Publishing Corp. (1946).
4. F.A. Cotton and G. Wilkinson, *Advanced Inorganic Chemistry*, 4th Ed., John Wiley & Sons, New York (1980).
5. *Chemical & Engineering News*, Oct. 11, 1982, p. 29.
6. L.F. Coleman, *Chemical & Engineering News*, Nov. 22, 1982, p. 2.
7. C.B. Allen and G.J. Janz, *J. Hazard. Mat.* **4**, 145 (1980).
8. G.D. Muir (Ed.), *Hazards in the Chemical Laboratory*, The Chemical Society, London (1977).
9. M.E. Green and A. Turk, *Safety in Working with Chemicals*, Macmillan Publ. Co., New York (1978).
10. The General Safety Committee of the Manufacturing Chemists' Assoc. Inc., *Guide for Safety in the Chemical Laboratory*, D. Van Nostrand, Inc., New York (1954).
11. Jon Eklund, *The Incompleat Chymist*, Smithsonian Institute Press, City of Washington (1975).
12. Georg Lockemann, *Selected Readings in the History of Chemistry*, A.J. Ihde and W.F. Kieffer (Eds.), Journal of Chemical Education Publ. (1965), p. 78.
13. H.A. Laitinen and G.W. Ewing (Eds.), *A History of Analytical Chemistry*, American Chemical Society, Washington, D.C. (1977), pp. 44–52.

14. R.A. Bailey and G.J. Janz, *The Chemistry of Non-Aqueous Solvents*, J.J. Lagowski (Ed.), Chapter 7, Academic Press, New York (1966).
15. I.E. Campbell, *High Temperature Technology*, John Wiley & Sons, New York (1956).
16. J.L. Margrave and R.H. Hauge, "Chemical Experimentation Under Extreme Conditions," Vol. IX of *Techniques of Chemistry Series*, B.W. Rossiter (Ed.), Chapter VI, Wiley-Interscience, New York (1980).
17. D.F. Shriver, *The Manipulation of Air-Sensitive Compounds*, McGraw-Hill Book Company, New York (1969).
18. E.S. Perry and A. Weissberger (Eds.), *Laboratory Engineering and Manipulations*, 3rd Ed., in *Techniques of Chemistry Series*, Vol. XIII, Wiley-Interscience, New York (1979).
19. J.D. Heldman, *Techniques of Glass Manipulation in Scientific Research*, Prentice-Hall, Inc., New York (1946).
20. R. Barbour, *Glassblowing of Laboratory Technicians*, 2nd. Ed., Pergamon Press, New York (1978).
21. J.L. White, in *Physico-Chemical Measurements at High Temperatures*, J.O'M. Bockris, J.L. White, and J.D. Mackenzie (Eds.), Chapter 8, Butterworths, London (1959).

2

Alkali Metal Carbonates

Hansraj C. Maru

1. Introduction

Among the carbonate melts, the class of alkali metal carbonates (Li_2CO_3, Na_2CO_3, and K_2CO_3) and their mixtures are the most prominent. These melts are fairly stable in the temperature range of 400 to 1000°C, exhibit low vapor pressures, can be easily contained in common ceramic and alloy materials, and are environmentally safe. The alkali carbonate melts offer a wide range of physicochemical and kinetic characteristics desirable for various reaction media and electrolytes. Melting points of mixtures vary from 400 to 900°C. Other properties, such as viscosity, electrical conductivity, and gas solubility can also be varied by changing the compositions of the mixtures. Applications[1] of carbonate melts include fuel cells, thermal energy storage, coal gasification, desulfurization of gases, oxygen regeneration, carbon dioxide concentration, nuclear and toxic waste processing, and recovery of silver. A list of some of these applications, their temperature range of interest, and the carbonate mixtures employed is presented in Table I. Preparation, analysis, and handling of these melts for research, development, and commercial use in these applications is therefore of interest.

Thus, the remaining text is divided into the following sections:

Section 2 (Preparation) examines the purity levels of commercial alkali carbonates and suggests techniques for the purification of melts. Mixing, grinding, and sieving operations required during the preparation of melts are also discussed.

Section 3 (Analysis) presents chemical analysis and measurement of im-

Hansraj C. Maru • Energy Research Corporation, 3 Great Pasture Road, Danbury, Connecticut 06810, U.S.A.

TABLE 1. Applications of Molten Alkali Carbonates

Application	Temperature range, °C	Carbonate mixture
Fuel cell	600–700	Li/K, Li/Na and Li/Na/K mixtures
Thermal energy storage	400–600	Li/K, Li/Na and Li/Na/K mixtures
Coal gasification	800–1000	Na_2CO_3
Desulfurization of gases	400–700	Li/Na/K eutectic
Nuclear and toxic waste processing	850–950	Na_2CO_3
Oxygen regeneration and CO_2 removal	600–700	Li/K or Li/K-CO_3^{2-}/Cl^- mixtures
Recovery of Ag from photographic films	850–950	Na_2Co_3

portant physicochemical properties, with emphasis on specific techniques suitable for molten carbonates.

Section 4 (Handling) recommends materials required for containment, electrochemical cell components, and hardware. Safety considerations are also briefly discussed in this section.

Some aspects common to all molten salt systems are dealt with in other chapters and therefore not duplicated here.

2. *Preparation*

2.1. *Purity Levels*

The specifications for the purity levels of individual carbonates will clearly depend on the intended end use. For example, in electrochemical and corrosion measurements, very pure oxide- and hydroxide-free melts may be desired. A small amount of impurities may be tolerated in certain physical property measurements, but some impurities may yield misleading results in other physical property measurements.

An interesting example of the above was reported by Eberhart.[2] When Li_2CO_3–K_2CO_3 mixtures made with ultrapure Li_2CO_3 (99.999% pure, Apache Chemicals, Benson, AZ, USA) and ultrapure or Reagent Grade K_2CO_3 were melted, they displayed sharp melting and complete wetting of nickel in a reducing gas environment. However, when a Reagent Grade Li_2CO_3 (~99.1% pure, Fisher Scientific, Fairlawn, NJ, USA) was used instead of the ultrapure grade, the mixtures displayed anomalous melting and wetting behavior. The initial melting point shifted upward by 50 to 100°C, and no wetting on nickel or stainless steel surfaces was observed. A surface-active impurity leading to the inhibition of wetting was suspected in the Reagent Grade Li_2CO_3. Note that both the Reagent Grade and the ultrapure materials displayed similar wetting behavior in an oxidizing environment. Thus appropriate purity levels are important in fundamental work.

In commercial use, a certain amount of oxide and hydroxide content may be tolerated (or may even exist because of an equilibrium with the environment), but other anions such as sulfates and halides may be deleterious from the containment and corrosion point of view.

Purity levels of the alkali carbonates vary with the source of the material, i.e., the raw material and the process used for preparation. A comparison of the typical specifications of these carbonates showed that technical grade carbonates are generally >99% pure (Tables II–IV). The sources of these materials and variations in purity levels and costs are briefly discussed in Sections 2.1.1–2.1.3.

2.1.1. Lithium Carbonate (Li_2CO_3)

Li_2CO_3 is obtained industrially by the reaction of soda ash with a solution of a soluble lithium salt, e.g., the sulfate or chloride. A large portion of Technical Grade Li_2CO_3 is used in aluminum production processes and ceramics. A special grade of high purity Li_2CO_3 is used in the treatment of manic-depressive psychoses. Because the chloride or sulfate is used as a starting material, the Technical Grade Li_2CO_3 may contain as much as 0.05% Cl^- and 0.3% SO_4^- impurities (Table II). The Purified Grade significantly reduces these impurity levels to 0.001 to 0.003%, although its price is three times that of the Technical Grade ($8.75/

TABLE II. Typical Analyses of Various Grades of Li_2CO_3 Powder

	Ultrapure	Reagent (%)	Technical (%)	Purified (%)
Assay (as Li_2CO_3)	99.985%	Min 99.0	99 to 99.2	99.4
H_2O (total)			0.5	0.5
Chloride (Cl)	<1 ppm	0.005	0.005–0.05	0.001
Sulfur compounds (as SO_4)	<1 ppm	0.2	0.15–0.3	0.003
Calcium (Ca)	20 ppm	0.01	0.04	0.01
Iron (Fe)	5 ppm	0.002	0.002	0.0004
Potassium (K)	2 ppm	0.005	0.02	0.001
Sodium (Na)	2 ppm	0.01	0.1	0.001
Magnesium (Mg)	0.1 ppm		0.01	
Heavy metal (as Pb)	~1 ppm	0.002		
Approximate price, first Q, 1981, \$/kg (quantity)	500–700 (0.1 to 1 kg)	35 to 50 (1 to 10 kg)	3.10 (up to 1000 kg)	11 (<50 kg)
		24 (truckload)	2.65 (truckload)	8.75 (truckload)

TABLE III. Typical Analyses of Various Grades of Na_2CO_3

	Ultrapure	Reagent	Synthetic ash	Natural ash
Assay (as Na_2CO_3)	99.98%	99.5%	99.8%	99.9%
Chloride (Cl)	3 ppm	0.002%	0.2%	0.01–0.02%
Sulfur compounds (as SO_4)	<1 ppm	0.003%	0.02%	0.02–0.03%
Calcium (Ca)	20 ppm	0.015%		
Iron (Fe)	0.1 ppm	5 ppm	15 ppm	7 ppm
Lithium (Li)	10 ppm			
Potassium (K)	1 ppm	0.005%		
Magnesium (Mg)	15 ppm	0.015%		
Heavy metals (as Pb)	~1 ppm	5 ppm		
Approximate price, first Q, 1981, $/kg (quantity)	1300 (100g)	12 to 20 (1 to 10 kg) 3.10 (truckload)	0.12 (bags, truckload) 0.09 (bulk, truckload)	

kg for Purified Grade against $2.65/kg for Technical Grade). The Purified Grade is also purer than the Reagent Grade (99.4% compared with 99.0% purity), although the latter is in fact three times more expensive ($24/kg for Reagent Grade against $8.75/kg for Purified Grade).

2.1.2. *Sodium Carbonate (Na_2CO_3)*

Na_2CO_3 is manufactured by a synthetic route (Solvay process) involving NaCl and $CaCO_3$ as the raw materials, or from naturally occurring trona (sodium

TABLE IV. Typical Analyses of Various Grades of K_2CO_3 Powder

	Ultrapure	Reagent	Technical
Assay (as K_2CO_3, after drying at 285°C)	99.93%	99.8%	99.0–99.6%
Halides (as Cl)	3 ppm	0.003%	0.02%
Sulfur compounds (as SO_4)	<1 ppm	0.004%	5 ppm
Calcium (Ca)	—	0.01%	0.003%
Iron (Fe)	0.02 ppm	5 ppm	3 ppm
Lithium (Li)	—		
Sodium (Na)	—	0.02%	0.2%
Magnesium (Mg)	—		0.001%
Heavy metals (as Pb)	—	<5 ppm	2 ppm
Silica (SiO_2)	—	0.005%	40 ppm
Approximate price, first Q, 1981, $/kg (quantity)	425 (100g)	20–25 (1 to 10 kg) 5 (truckload)	0.65 (truckload)

sesquicarbonate, $Na_2CO_3 \cdot NaHCO_3 \cdot 2H_2O$) deposits. Chloride and calcium may therefore be expected to be the main impurities in the Synthetic Ash. The Natural Ash obtained from trona, however, contains very little chloride (0.01 to 0.02%) because chloride is absent in the raw material, trona. The price of both the products, however, is comparable ($0.09/kg). Thus, if available the Natural Ash may be preferable to Synthetic Ash. The actual specifications of the Na_2CO_3 product or the supplier may be consulted to obtain the desired grade.

2.1.3. Potassium Carbonate (K_2CO_3)

K_2CO_3 is generally manufactured by electrolysis of KCl to produce KOH, followed by CO_2 absorption to produce K_2CO_3. The carbonate is usually purified by washing after crystallization, or by recrystallization. Because KCl is used as a raw material, chloride is expected to be the main impurity, but its extent is small compared with Synthetic Ash Grade Na_2CO_3. Many potassium compounds are used in medicine to such an extent that essentially all the potassium compounds in the United States are manufactured to pass the U.S. Pharmacopoeia (USP) tests for purity. Thus Technical Grade K_2CO_3 has only 0.02% halide impurities and 40 ppm SiO_2.

Overall, the Technical Grade K_2CO_3 (at $0.65/kg) is approximately seven times more expensive than the Technical Grade Na_2CO_3 (at $0.09/kg), but four times cheaper than the Technical Grade Li_2CO_3 (at $2.65/kg). Thus, if the cost is a main consideration in an application, mixtures containing the maximum amount of Na_2CO_3 should be used.

2.2. Mixing, Grinding, and Sieving

When a mixture of carbonates is required for an application, proper premixing is recommended. Because the alkali carbonates are hygroscopic, lump-free mixing will require handling in a low humidity environment. Establishing a controlled humidity room is recommended for this purpose, particularly if carbonates are processed routinely. A typical room may consist of specially moisture-proofed walls and be equipped with an absorption dryer (e.g., Cargocaire Dehumidification Systems, Amesbury, MA, USA). An air conditioner is essential to maintain a comfortable temperature for personnel and also to help reduce the humidity level. A double door arrangement will minimize the infiltration of outside air. An absolute zero humidity is neither essential nor practical. We have found that a "dry room" with about 10 to 15% relative humidity provides an adequate atmosphere for most operations. If, for any reason, a very dry atmosphere is desired, a desiccator may be kept inside the dry room. If it is not practical to set up a dry room, a desiccator or properly sealed dry box may also be used. For mixing and grinding outside a dry room, moisture-tight caps should be provided for the jars.

The material of the inside wall of a mixer or grinder should be properly selected, depending on the purity requirements for the end use. For fuel cell applications, we have found high density alumina jars and balls adequate. The jars should be thoroughly dried before use.

Sieving carbonate powders is difficult, particularly if very fine mesh (>100 mesh) screens are used. The use of a dry room and a light brush to aid in sieving are helpful. In many cases, a "sieve analyzed" product is adequate. In this procedure, grinding and sieve analysis are carried out in sequence but, rather than sieving the whole batch, only a sample is sieved to determine if the desired particle size distribution has been attained.

An intimate mixing of various carbonates may be obtained by premelting them together. The powder can be loaded in an aluminum or high purity alumina crucible. To minimize the reaction of carbonates with the crucible walls (leading to formation of alkali metal aluminates), preconditioning of the inside surface of the crucible with carbonate melt is recommended. A further discussion of this is presented in Section 4. The furnace atmosphere should be maintained at >5 to 10% CO_2 to prevent decomposition of carbonates. After melting the mixture at approximately 50°C above the melting point for an hour, the melt may be poured into a shallow aluminum pan in a thin layer. The thin layer of the melt poured in the pan solidifies immediately and does not stick to the pan. The sheet can be broken up easily with a pestle, ground to a desired particle size in a ball mill and then "sieve analyzed" as discussed above. Care should be exercised not to leave behind any solidified material in the crucible. Depending on the composition of the melt, it is likely that the composition of the solidified residue may be different from the "average" intended mixture. To avoid this, a remelting step may also be added. The ultimate check, of course, is to analyze the resultant mixture. We have found it best to avoid all premelting steps.

2.3. *Purification*

Because the alkali carbonates are hygroscopic, some hydroxide is normally present in the salts. Often, some oxides may also be present. The purification procedure used by Andersen[3] for his thermodynamic measurements appears safe for conversion of these oxides and hydroxides to carbonates:

1. Dry the carbonates for a period of 24 h or more at about 200°C.
2. Heat the salt or salt mixture slowly and melt it in a dry CO_2 atmosphere.
3. Bubble dry CO_2 through the melt for at least 12 h.

The results of hydrolysis constant measurements confirmed the validity of this procedure. Appleby and Nicholson[4] have used a somewhat more rigorous method for their electrochemical measurements. After melting under CO_2, they pumped down the melt to the CO_2 decomposition pressure for degassing and removal of water vapor. The pressure was reduced slowly, over several hours, to this level. The melt was then saturated by bubbling CO_2 to neutralize any O^{2-} ions formed and to decompose any OH^- ions present. Carbon dioxide used for this purpose should be of high purity and should be dried over a silica gel or molecular sieve absorber.

If chloride and sulfate impurities are of concern in a particular application, starting materials containing low amounts of these impurities should be chosen whenever practical. In a molten carbonate fuel cell, the chloride impurities are purged easily because of their higher vapor pressure compared with carbonate salts. For laboratory investigations, the purification process described above may also clean up other impurities in addition to ensuring the conversion of hydroxides and oxides to carbonates.

3. *Analysis*

In this section, we will discuss the analysis of cations, carbonates, oxides, and hydroxide impurities. Some discussion of the measurement of physical properties relevant to carbonates will also be presented. Many of the standard analytical and measurement techniques, especially those utilized for other molten salts, are applicable to molten carbonates as well, so the emphasis in this section will be on special techniques adopted by workers in the carbonate field and on the precautions required.

3.1. *Chemical Analysis*

One of the most important precautions during the analysis of carbonates is to avoid any weight changes resulting from the absorption of moisture, which can change total weights as well as the composition of the sample under investigation.

The composition of various alkalies is most easily obtained by atomic absorption spectrometry. A total carbonate content can be obtained by measuring the CO_2 evolved by reaction with HCl or H_2SO_4. Oxides and hydroxides associated with alkalies can be estimated by a material balance, using the information derived from the atomic absorption and CO_2 analyses. Alternatively,[3] the sample can be dissolved in water; the carbonates are then precipitated out by the addition of $BaCl_2$, and the OH^- content is determined by pH-metric titration. Note that Li_2CO_3 is only sparingly soluble in water, its solubility decreasing as the temperature increases. K_2CO_3 and Na_2CO_3 solubilities on the other hand increase as the temperature increases. Enough water should therefore be added for a proper dissolution of all the salts.

In molten carbonate fuel cell applications, where $LiAlO_2$ or some other alkali containing mixed oxides is used as an electrolyte support material, a "washing" procedure is required for a clean separation of the carbonates from the support material. Because $LiAlO_2$ hydrolyzes in the presence of water, water cannot be used as a solvent for leaching the carbonates. A 70% glacial acetic acid/30% acetic anhydride solution is quite adequate for this purpose.[5] The 30% acetic anhydride is added to minimize the formation of aluminum acetate ($Al[OH][CH_3COO]_2$).

If an equilibrium composition of a melt in contact with a given hot gas environment is desired, the composition needs to be "frozen" before analysis. This may be accomplished by drawing a sample from the melt and blowing it onto a cold surface.

3.2. *Measurement of Physicochemical Properties*

Extensive data on various physical properties, such as solid–liquid phase equilibria, density, surface tension, viscosity, and electrical conductivity are available[1,6,7] for mixtures of alkali carbonates and, in some cases, for mixtures with other additives. These properties for individual carbonates and some common salt mixtures at 700°C are summarized in Table V. The literature data may be sufficient in most cases, but if measurements are needed under special conditions or in the presence of certain additives or impurities, it is recommended that the procedures and precautions outlined below be followed. Also, because data on vapor pressure, solubility of gases, and diffusivities of dissolved species are limited, these measurements are discussed in greater detail.

Solid-Liquid Phase Equilibria may be determined by the conventional cooling curve method. "Seeding" with a small amount of solid carbonate may be used to avoid "supercooling." Special equipment modifications made by Cairns and MacDonald[8] may be employed if a highly sensitive analysis is desired. A phase diagram for the ternary system of carbonates[9] is presented in Figure 1. The compositions of various eutectic mixtures and corresponding melting points are presented in Table V. Compositions and the melting points are judged to be the best estimates from the data reported in the literature to date.

Density and Surface Tension may be measured simultaneously by using a double cone bob[10] and employing the Archimedian principle and the duNouy tensiometer theory. Such an apparatus, along with the detail of the bob, is shown in Figure 2. As with any surface tension measurements, the effects of impurities, gas atmosphere, and surface characteristics of the bob should be kept in mind.

Viscosity may be measured by the principle of the oscillating crucible, wherein a logarithmic decay of the oscillations is measured to determine viscosity.[11] The apparatus based on this principle is shown in Figure 3.

Electrical Conductivity may be measured in a capillary-type conductance cell as shown in Figure 4. A paper by Braunstein and Robbins also provides a comprehensive review of the subject.[12]

Vapor Pressure is an indicator of the stability of the melt. For carbonate melts, several vapor-liquid equilibria are possible[13] as shown in Figure 5. The simplest of them, the equilibrium dissociation pressure is due to the reaction

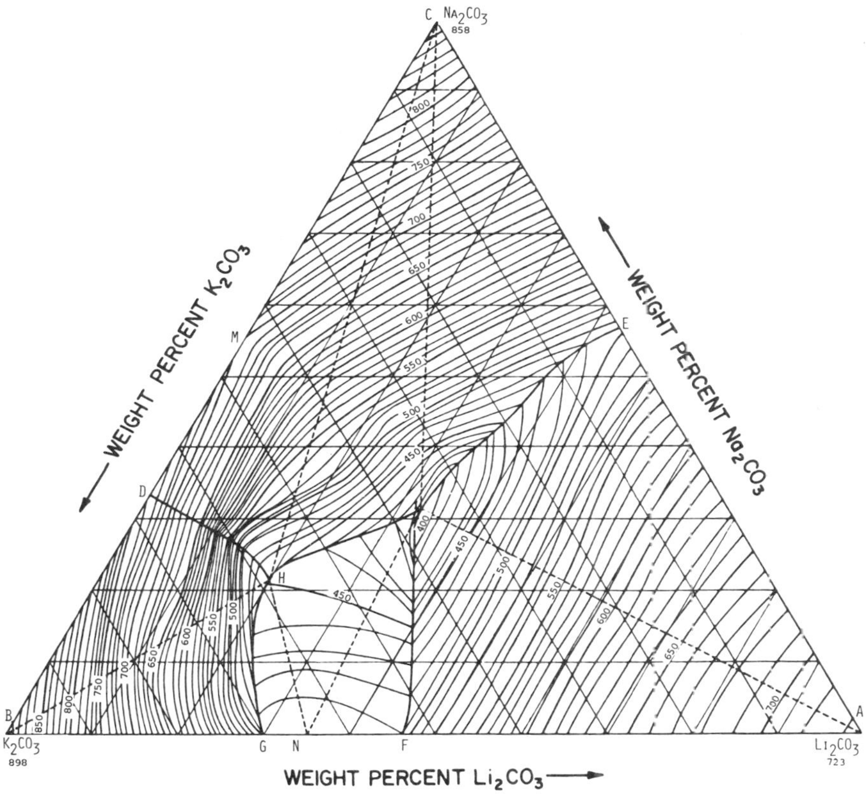

FIGURE 1. Phase Diagram of the Ternary System Li_2CO_3–Na_2CO_3–K_2CO_3.[9] (Temperatures in °C.)

$$M_2CO_3(l) \rightleftharpoons M_2O(l) + CO_2(g) \tag{1}$$

Other equilibria involve a variety of gaseous alkali metal species, including $M(g)$, $MOH(g)$, $M_2(OH)_2$, $M_2CO_3(g)$, $M_2O(g)$, and $M_2O_2(g)$, as shown in Figure 5. Approximate equilibrium constants for appropriate reactions are also shown in this figure.

The term vapor pressure has been used in the molten carbonate literature to indicate the irreversible loss of alkali species in a flowing gas environment. An estimate of this irreversible loss under different gas environments is best obtained by a transpiration method.[14] In this method, the vapors generated by the sample are carried away by a stream of a carrier gas continuously flowing over the sample. An alumina crucible containing the melt is suspended by a gold wire from a high accuracy balance (e.g., a Mettler five-digit balance with a bottom-loading provision or a Cahn Balance). Figure 6 shows one such arrange-

TABLE V. Physicochemical Properties of Alkali Carbonate Mixtures

Salt/salt mixture[a]	Composition[b]						Melting point, °C	Approximate value of property at 700°C[c]						
	Mol %			Wt %				Density, kg/m^3	Surface tension, mN/m	Viscosity, mN-s/m^2	Electrical conductivity, $(\Omega\text{-cm})^{-1}$	Vapor loss,[d] g/s cm^3 $\times 10^{-10}$	Solubility, mol/cm^3-atm $\times 10^{-7}$	
	A	B	C	A	B	C							O_2	CO
Li_2CO_3	100	—	—	100	—	—	723	(1843)	(245)					
Na_2CO_3	—	100	—	—	100	—	858	(2038)	(220)					
K_2CO_3	—	—	100	—	—	100	898	(1988)	(182)					
$LiNaCO_3$(c)	50.3	49.7	—	41.4	58.6	—	501			6.5				
Li_2CO_3-Na_2CO_3(e)	53.3	46.7	—	44.3	55.7	—	500	1937	239	6.55	2.5	5	2.1	2.2
Li_2CO_3-K_2CO_3(el)	44.5	—	55.5	30.0	—	70.0	500	1921	203	5.3	1.3		4.2	2.4
$LiKCO_3$(c)	50	—	50	34.8	—	65.2	505	1917	206	5.45				
Li_2CO_3-K_2CO_3(e2)	62	—	38	46.6	—	53.4	488	1912	214	5.7		25		
Na_2CO_3-K_2CO_3(l)	—	56	44	—	49.4	50.6	710	(2013)	(214)				(6.4)	(5.4)
Li_2CO_3-Na_2CO_3-K_2CO_3(e)	44.1	29.8	26.1	32.5	31.5	36.0	393	1984	220		1.7	25	2.5	0.9

[a] c = compound, e = eutectic, l = low melting liquid.
[b] A = Li_2CO_3, B = Na_2CO_3, C = K_2CO_3.
[c] Parentheses indicate values extrapolated assuming supercooled liquids.
[d] g carbonate lost from the melt per standard cm^3 of 14 O_2/28 CO_2/51 N_2/7 H_2O gas.

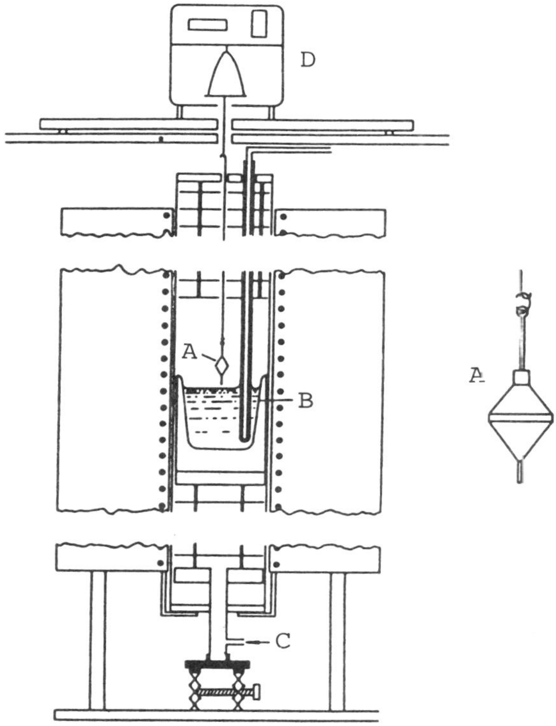

FIGURE 2. Apparatus for Density–Surface Tension Measurement.[10] (*A:* Bob; *B:* Crucible with Molten Salt; *C:* Variable Height Mount with CO_2 Gas Inlet; *D:* Analytical Balance.) Reprinted by permission of the Publisher, The Electrochemical Society.

ment used at the Institute of Gas Technology.[13] The weight measurements are taken periodically in this case by removing Plug C and suspending the crucible from the balance. The plug prevents any back diffusion of outside gases. The gases flow downward in the quartz tube which houses the alumina crucible. The downward flow prevents any undesirable condensation of vapor on the suspension wire. A bubbler at the exit provides a slight positive pressure in the quartz tube. The entire apparatus can be enclosed in a pressure vessel to allow measurements under pressure or vacuum.

Separate runs with different flow rates of the carrier gases should be made to check whether the flow rates used are on the flat portion of the "S" curve (apparent vapor pressure against flow rate).[14]

It may be noted that this transpiration method can be used to estimate several vapor–liquid equilibria separately, if so desired. For example, the dissociation pressure may be determined by using dry CO_2-free air as the carrier gas, and

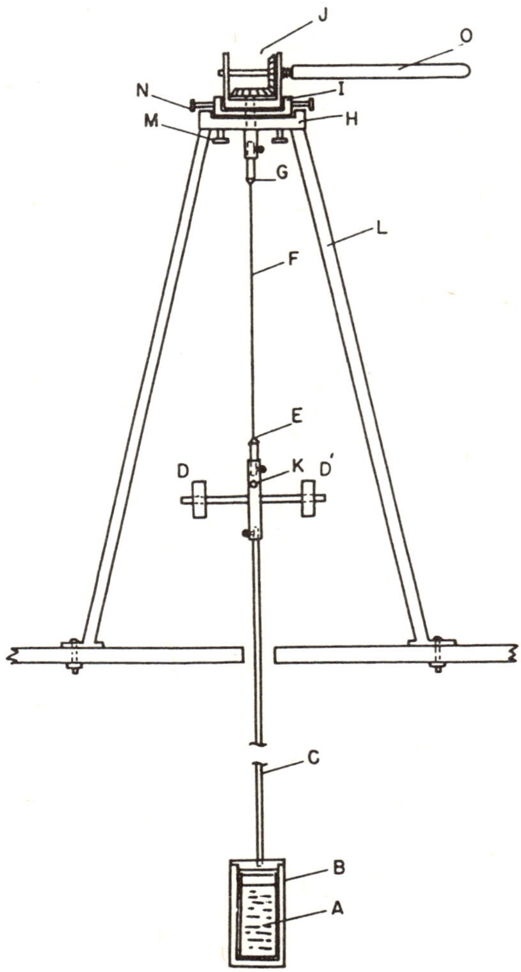

FIGURE 3. Cross-sectional view of the Torsion Head Suspension System and crucible of the "Oscillating" Viscometer.[11] (*A:* Cylindrical Crucible, 80%–20% Au–Pd; *B:* Inconel Cradel; *C:* Nickel Rod, 6 mm in Diameter and 560 mm Long; *DD':* Inertia Piece, 25 mm and 12 mm Thick; *E:* Pinchuck; *F:* Tungsten Suspension Wire, 40 s.w.g.; *G:* Pinchuck; *H:* Torsion Head; *I:* Aluminum Base-Plate; *J:* Gear Assembly; *K:* Mirror; *L:* Tripod; *M, N,:* Screws; *O:* Aluminum Rod.) Reprinted by permission of the Publisher, The Electrochemical Society.

noting the weight loss and analyzing the gas for CO_2. Similarly, other equilibria may be measured by suitably modifying the carrier gas with different partial pressures of O_2, H_2, H_2O, etc. Vapor loss data obtained by IGT in a typical oxidant gas environment are shown in Table V. The measurements are reported

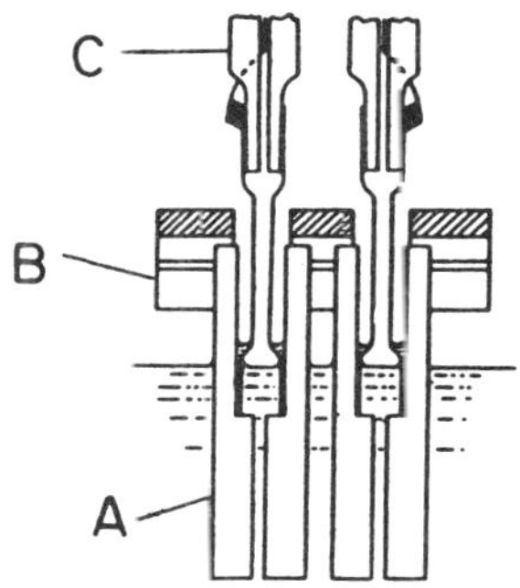

FIGURE 4. Conductance cell design.[10] (*A:* Single Crystel MgO Capillaries; *B:* Au–20% Pd Block; *C:* Insulated Electrode Leads.) Reprinted by permission of the Publisher, The Electrochemical Society.

as grams of weight loss from the melt for a standard cubic centimeter of gas flowing over the melt. Although these losses appear to be small, when several million liters of gas pass through a fuel cell during its life, significant losses may be encountered. Therefore, a salt mixture with low vapor pressure is preferable. The Li_2CO_2–Na_2CO_3 mixture has shown vapor losses five times lower than the Li_2CO_3–K_2CO_3 or Li_2CO_3–Na_2CO_3–K_2CO_3 eutectics (Table V) so, if vapor loss is an important consideration, a Li_2CO_3–Na_2CO_3 mixture may be preferable.

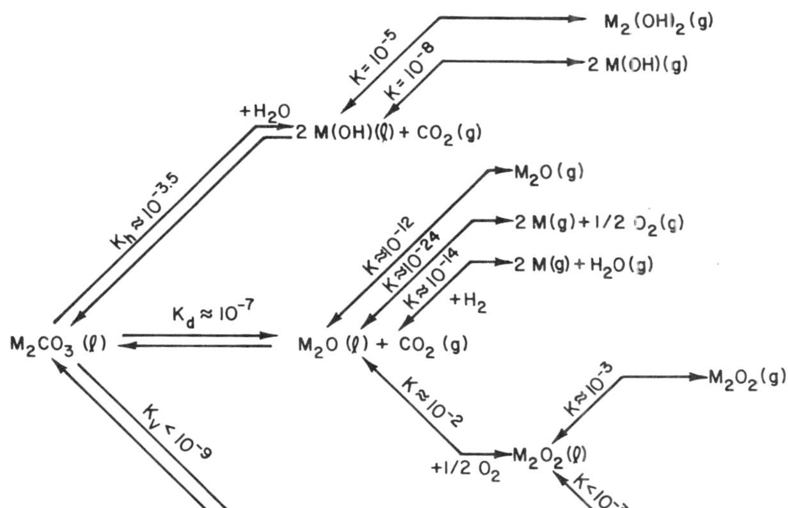

FIGURE 5. Reaction equilibria and possible volatilization mechanisms.[13] (K values typical of low-melting eutectic carbonates at 700°C, estimated using ideal mixture rule.)

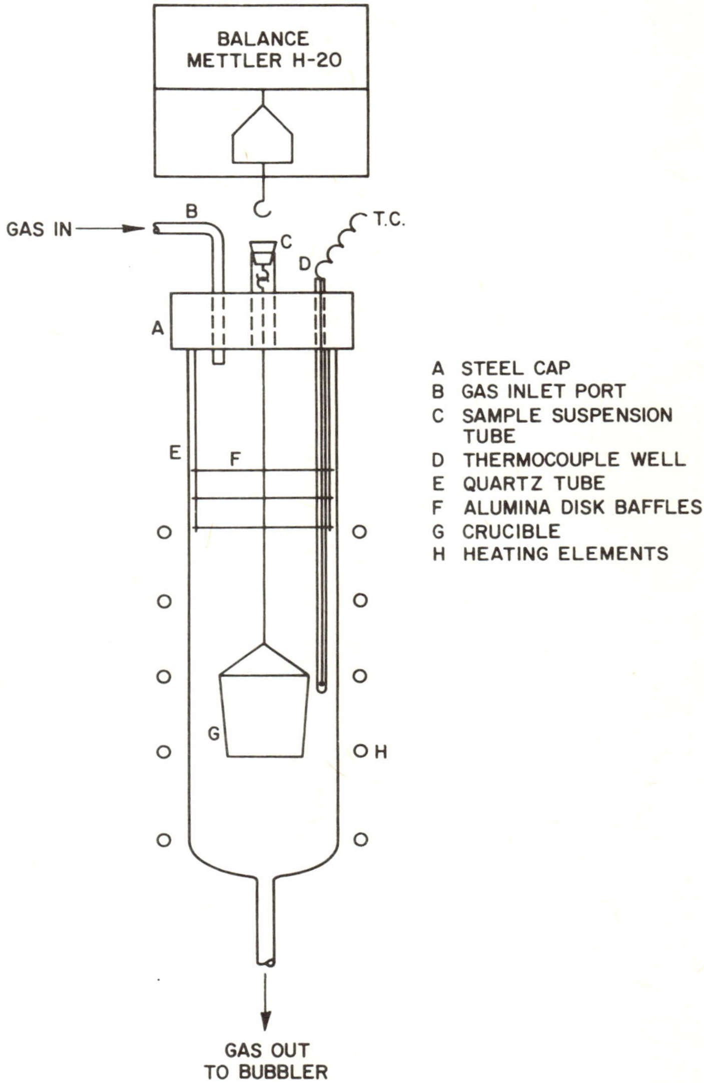

FIGURE 6. Transpiration apparatus for vapor pressure measurements.[13]

Solubility of Gases in carbonate melts is an important factor in many applications. For unambiguous determination of solubility, care should be exercised to define the solubility properly and to maintain an appropriate gas composition. Broers and co-workers[15] have used amperometric titration for the determination of solubility of H_2, CO, CH_4, and O_2. Powdered $K_2Cr_2O_7$ was used as a titrant for H_2, CO, and CH_4, whereas the addition of excess $Cr_2(SO_4)_3$ followed by chromate was used for O_2 solubility measurements. Appleby and Van Drunen[16] have used a quenching or melt chilling method for solubility measurements. It measures the quantity of dissolved gas released on freezing of the melt. The advantages of this method include rapid measurement and the possibility of determining solubilities of several gases simultaneously. Their apparatus (Figure 7) consists of a silica or quartz tube connected to a vacuum. The carbonate melt

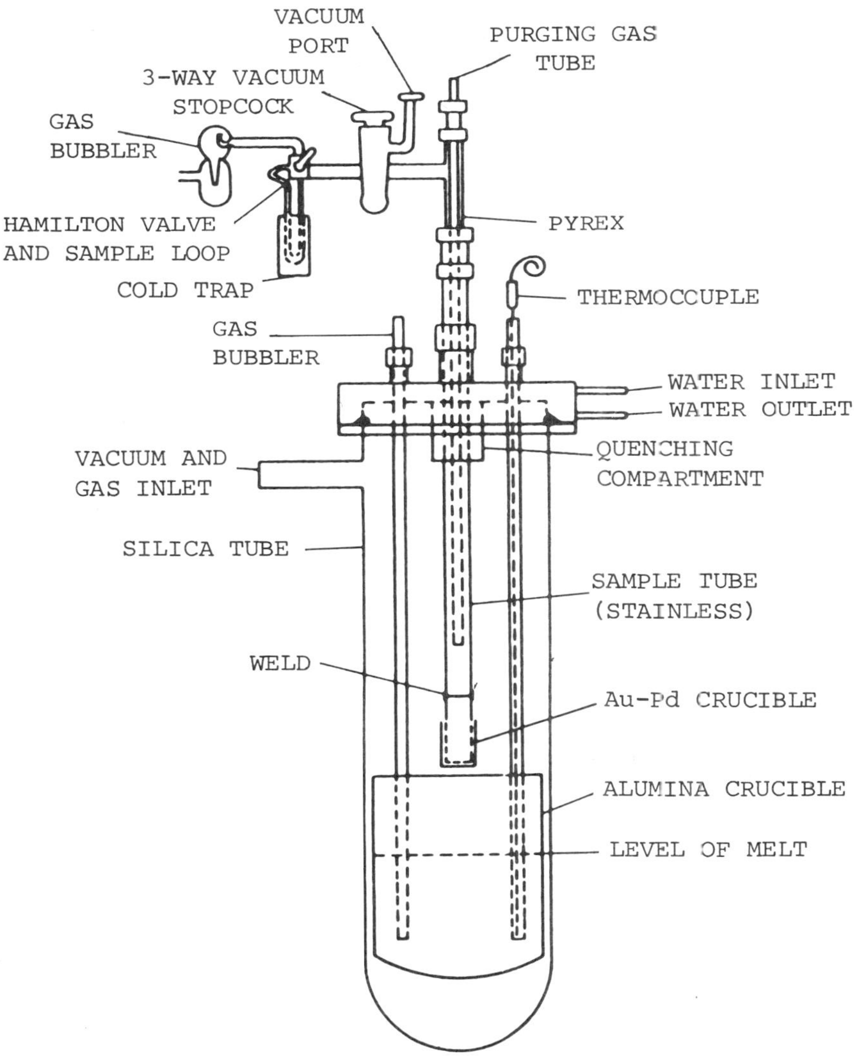

FIGURE 7. Apparatus for solubility measurements.[16] Reprinted by permission of the Publisher, The Electrochemical Society.

is placed in an alumina crucible and saturated by the desired gases using a bubbler. After saturation for 2 to 3 h, a sample is withdrawn in a small Au–Pd crucible and introduced into the upper quenching compartment by vertical translation of the support tube in its sliding seal attached to the water-cooled stainless steel header. On slow quenching, the gas is liberated, flushed by a stream of helium, and then trapped in a U-tube containing an activated Linde 5 Å molecular sieve. After being allowed to reach room temperature, the entire contents of the tube are flushed with a stream of carrier gas into a gas partitioner and analyzed.

The solubilities of most gases of interest may be determined by this method. Where gas equilibria result in a different composition (e.g., H_2 and CO_2 equilibrate by the water–gas shift reaction), separate determinations using more than one gas mixture may be required to obtain estimates of the solubilities of individual gases.

Diffusivities of Dissolved Gases in the melt may be determined by chronopotentiometric measurements, as was done by Volgin *et al.*[17] The measurements yield $cD^{1/2}$, where c is the solubility of the gas under investigation. Although Volgin's measurements suffered from an inaccurate determination of c, appropriate determination of solubility as described in previous subsections may alleviate this problem. Volgin reported a diffusivity of H_2 at 700°C as 6.2×10^{-4} cm^2/s for the Li_2CO_3–Na_2CO_3 eutectic. This value, however, appears somewhat high, because his solubility values are somewhat questionable. If one uses the Wilke-Chang equation, the predicted values would be an order of magnitude smaller than those reported by Volgin.

3.3. Electrochemical Measurements

Most of the electrochemical measurement techniques, particularly those employed for molten salt systems, are applicable to carbonates. A selection of appropriate working and reference electrodes and container materials is, of course, important. A discussion of the container materials is presented in Section 4 and a discussion of reference electrodes is presented below.

The reference electrodes used in the past for carbonate applications include[1] metal/metal ion electrodes like Ag/Ag^+, Pb^{2+}/Pb, Cd^{2+}/Cd, and Bi^{3+}/Bi, and gas reference electrodes O_2, CO_2/CO_3^{2-}, and CO_2/CO, CO_3^{2-}. Because of its excellent stability and reversibility, the O_2, CO_2/CO_3^{2-} reference electrode has become one of the most popular reference electrodes for molten carbonate measurements.

This electrode is based on the overall reaction

$$CO_3^{2-} \rightleftharpoons CO_2 + 1/2\ O_2 + 2e^- \qquad (2)$$

A typical reference electrode construction for fuel cell applications[18] is shown in Figure 8a. It consists of an outer alumina tube of approximately 6 mm outside diameter and with a fine ~0.5 mm diameter hole (E) at the bottom. The tube

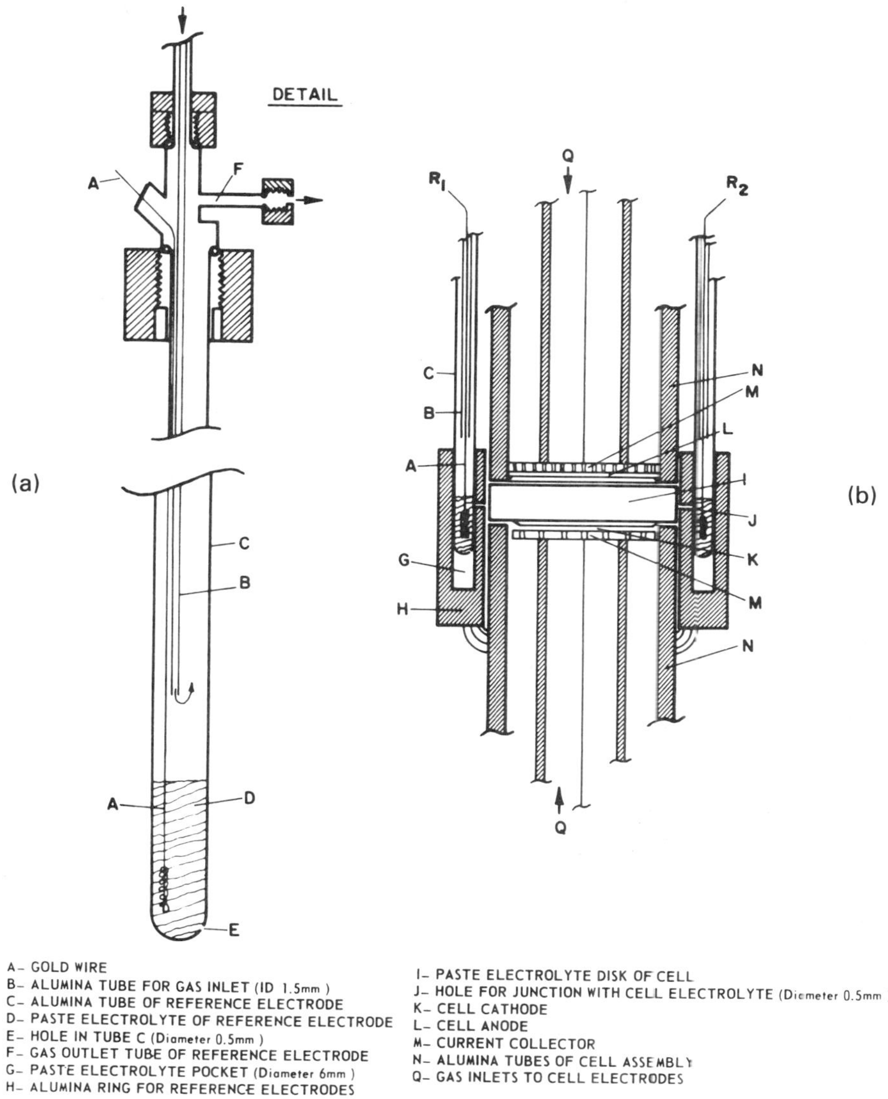

FIGURE 8. Reference Electrode Assembly: a) Reference tube; b) Dual reference electrode system for a fuel cell.[18] Reprinted by permission of the Publisher, The Electrochemical Society.

is filled with a molten-carbonate/inert support (e.g., $LiAlO_2$) mixture and a pure gold wire (0.5 mm diameter) is immersed in the melt. An inner alumina tube of about 2 mm outer diameter situated a few centimeters above the electrolyte level supplies the reference gas mixture, $P_{CO_2} = 2/3$ atm and $P_{O_2} = 1/3$ atm.

Two of these reference electrodes are shown implemented in a 3 cm^2 fuel cell (Figure 8b). Here the outer sleeve (H) (inside diameter = 25 mm) containing two pockets (G) of approximately 6 mm inside diameter houses the reference electrodes. Fine cross holes of 0.5 mm diameter (drilled from the outside) serve as junctions between the cell and the reference electrodes. The external holes (J) and the outer sleeve (H) are sealed from the outside with an alumina–phosphoric acid paste, followed by heating at 1000°C. The pockets are also filled with a carbonate/support mixture.

Some precautions should be exercised to obtain reliable results:

a. To avoid any inaccuracies in the Nernst potential, the reference gas mixture composition should be held constant as closely as possible at 2/3 CO_2 : 1/3 O_2.
b. The packing and composition of the carbonate/support mixture in the reference tube and the cavity should be as similar to the working electrolyte as possible. This will avoid the capillary and diffusion interchange of species between the reference and the working compartments.
c. The reference tube should be well supported and should be heated and cooled slowly to avoid cracking during experimentation.

An alternative to the O_2, CO_2/CO_3^{2-} gas reference electrode may be a CO_2/CO, CO_3^{2-} electrode. This electrode system has been preferred by some investigators,[19] particularly when the working-electrode gas is a reducing gas. Care must be exercised, however, to avoid those compositions which lead to carbon formation by the Boudouard equilibrium ($2\ CO \rightleftharpoons CO_2 + C$). Some inaccuracies may also arise if water vapor is present in the gas phase ($CO + H_2O \rightleftharpoons CO_2 + H_2$) or if the melt contains hydroxide. Hydrogen is formed in either case. Also, when using oxidizing working gases, some safety considerations are involved to avoid the accidental mixing with the CO_2/CO reference gas.

At high temperatures (~1000°C), Kim and Devereux[20] have reported that a McDanel Type 998 close-ended alumina tube of 13 mm outer diameter and a 1.6 mm wall thickness served as an adequate oxide ion junction for a CO_2/CO, CO_3^{2-} reference electrode. This avoids any direct contact of the reference and working sections. Effectiveness of the ionic junction at lower temperatures is, however, questionable. Braunstein *et al.*[21] have stated that a sodium β (or β″−) alumina may serve well at lower temperatures in melts containing Na_2CO_3.

For carbonate melts containing chlorides, a graphite, Cl_2/Cl^- reference may also be utilized.[1]

4. Handling

The two main considerations in handling of molten carbonates are materials of construction and safety.

4.1. Materials

Material requirements for molten carbonate applications include structural strength, corrosion resistance, high surface area for support and reactivity, high conductivity for current collection, high dielectric strength for insulation, etc. In some instances, these requirements (which are often contradictory) are imposed simultaneously. The stability and suitability of a particular material depend greatly on

a. The melt composition, including minor impurities and moisture content;
b. The surrounding gas environment, especially CO_2, H_2O, SO_2, and Cl_2;
c. The physical form of the material, such as powder or porous sinter, fibers, coatings, or bulk solid;
d. The operating potential, temperature and pressure.

Various metals, alloys, ceramics, semiconductors, coatings, and composite materials may be employed. A good discussion of the thermodynamics and corrosion behavior of different metals and ceramics is presented in Reference 1 and is not repeated here. Instead, an attempt will be made to recommend specific materials for containment, cell components, cell seals, and laboratory apparatus.

4.1.1. Containment Materials

A wide variety of materials may be utilized for the containment of carbonates. Significant experience has been obtained with the corrosion of different materials in various applications. Comprehensive data[22–24] on the corrosion of different materials for a variety of electrolyte compositions, gas environments, and temperatures are presented in Table VI.

The majority of the data published so far pertain to oxidizing environments. Generally, for most practical purposes, the effect of electrolyte composition (i.e., ratio of Li, Na, and K) may be small compared with the effect of gas environment, temperature, and potential. Thus at 500°C, Stainless Steel Type 304L, Type 1020 steel, and some types of graphite materials may provide adequate containment. At 600 to 700°C, a somewhat better corrosion resistance is required, and materials such as Stainless Steel Types 310, 316, 321, and 347 may be selected. Although it is not listed in the table, aluminum may also be used as a containment material for temperatures up to 600°C. Two advantages of using aluminum are its low cost and low contamination of the melt. A protective $LiAlO_2$ scale is formed on aluminum. Structural strength can be a problem as the temperature approaches the melting point of Al (660°C). Some alloys of aluminum may be used to improve the strength but they need to be investigated before they can be used. At higher temperatures, Kanthal-type materials (Fe-Cr-Al) containing 4 to 5.5% Al may be desirable. These materials form a protective alumina scale, which may convert to alkali aluminates in contact with the carbonates. At temperatures approaching 900 to 1200°C, only alumina appears to

TABLE VI. Corrosion Rates Observed by Various Investigators

Material	Composition of carbonates, mol percent	Gas environment	Temp., °C	Time, hr	Corrosion rate,[i] mil/yr	Remarks	Reference
1020 Steel	Ternary eutectic[e]	Air	500	50	4.8	Static test	22
Type 304 L SS	Ternary eutectic	Air	500	50	0.5	Static test	22
	Ternary eutectic	Air	600	50	−4.8[g]	Static test	22
Type 310 SS	52 Li_2CO_3 + 48 Na_2CO_3	Air + 1% CO_2	680	[f]	1.5	At −850 mV,[h] submerged	23
	52 Li_2CO_3 + 48 Na_2CO_3	Air + 1% CO_2	680	[f]	4.2	At −100 mV,[h] submerged	23
	Ternary eutectic	Air	600	50	4.8	Static test	22
	Ternary eutectic + 20% $S^=$	Air	600	50	185	Static test	22
Type 321 SS	52 Li_2CO_3 + 48 Na_2CO_3	Air + 1% CO_2	680	[f]	6.9	At −850 mV,[h] submerged	23
	52 Li_2CO_3 + 48 Na_2CO_3	Air + 1% CO_2	680	[f]	7.6	At −100 mV,[h] submerged	23
Type 347 SS	Ternary eutectic	CO_2	730	95	10	Submerged	24
	52 Li_2CO_3 + 48 Na_2CO_3	Air + 1% CO_2	680	[f]	2.2	At −850 mV,[h] submerged	23
	52 Li_2CO_3 + 48 Na_2CO_3	Air + 1% CO_2	680	[f]	1.3	At −100 mV,[h] submerged	23
	Ternary eutectic	Air	600	50	2.4	Static test	22
	Ternary eutectic	Air	700	50	3.1	Static test	22
	Ternary eutectic + 20% $S^=$	Air	600	50	440	Static test	22
	Ternary eutectic + 20% $S^=$	Air	700	50	100	Static test	22
	Ternary eutectic	Air	600	1500	0.3	Dynamic test	22
	Ternary eutectic	Air	700	1500	>90	Dynamic test	22
	Ternary eutectic	Air	500	8000	0.2	Dynamic test	22

Type 430 SS	Ternary eutectic	Air	500	50	2.9	Static test	22
Kanthal A	52 Li_2CO_3 + 48 Na_2CO_3	Air + 1% CO_2	680	[f]	Negligible	At −850 mV,[h] submerged	23
	52 Li_2CO_3 + 48 Na_2CO_3	Air + 1% CO_2	680	[f]	0.02	At −100 mV,[h] submerged	23
	Ternary eutectic + $LiAlO_2$	Air + 1% CO_2	700	1000	Negligible	Wires embedded in a paste mixture	23
Kanthal A-1	52 Li_2CO_3 + 48 Na_2CO_3	Air + 1% CO_2	680	[f]	Negligible	At −850 mV,[h] submerged	23
	52 Li_2CO_3 + 48 Na_2CO_3	Air + 1% CO_2	680	[f]	0.16	At −100 mV,[h] submerged	23
Uniloy (50 Cr/49Ni/1Ti)	Ternary eutectic + 20% $SO_4^=$	Air	700	1500	0.5–3	Dynamic test	22
	Ternary eutectic + 20% $S^=$	Air	700	1500	0.2–11	Dynamic test	22
	Ternary eutectic + 20% $SO_3^=$	Air	700	1500	1–5	Dynamic test	22
Graphite[a]	Ternary eutectic	Air	500	50	2.6	Static test	22
Graphite[b]	Ternary eutectic	Air	500	50	1.8	Static test	22
Graphite[c]	Ternary eutectic	Air	500	50	150	Static test	22
Graphite[d]	Ternary eutectic	Air	500	50	6.2	Static test	22
Alumina	Ternary eutectic	Air	500–900	50	Negligible	Static test	22

[a] AGSX graphite #571X.
[b] Carborundum Metal Pumping Service graphite.
[c] K.B. Bonite graphite.
[d] Raschig ring 571 B.
[e] Ternary eutectic composition is 43.5 mol% Li_2CO_3 + 31.5 mol% Na_2CO_3 + 25 mol% K_2CO_3.
[f] Potentiostatic polarization scans.
[g] Negative sign indicates a weight gain during the test.
[h] Working potential with reference to 67 CO_2 + 33 O_2 reference electrode.
[i] Multiply by 0.025 to convert mil/yr value to mm/yr.

be a cost effective container material. Alumina forms a protective layer of alkali aluminate by the reaction

$$Al_2O_3 + M_2CO_3 \rightleftharpoons 2MAlO_2 + CO_2; M = \text{alkali metal} \quad (3)$$

Note that because there is an increase in the volume of the material after reaction, a dense protective film of alkali aluminate is formed. A high density, high purity, fused-cast or slip-cast α-alumina should be used, because a low density or amorphous alumina is not corrosion resistant. Alumina with numerous voids or impurities, particularly silica, disintegrates in carbonate melts. Grantham and Ferry[25] have noted that the external crystal structure of cast alumina bricks is more corrosion resistant than the internal structure. Therefore care must be exercised in using such bricks.

Gold is highly resistant to corrosion in alkali carbonate melts and may be utilized if an application warrants it. For improved strength, some workers have utilized a Au-20% Pd (Palau) alloy, but Pd is attacked slowly by the melt, forming an alkali palladiate. Platinum is also attacked; silver even more rapidly.

If sulfur compounds or halides are present in the gas environment or in the melt, additional corrosion may be encountered. Grantham and Ferry note that sulfur compounds enhance corrosion, but chloride compounds do not alter the corrosion rate markedly. Among the sulfur compounds, sulfite is more corrosive than sulfide.

A reducing environment and water vapor present additional corrosion problems. These problems are discussed in Section 4.1.4. A discussion of electrode, electrolyte support, and hardware materials follows. Although this discussion is specific to molten carbonate fuel cells (Figure 9), the general principles should be similar for other applications.

4.1.2. *Electrode Materials*

Electrode materials for fuel cell applications are generally porous plaques with 50 to 80% porosity, surface areas ranging from 0.1 m^2/g to 1 m^2/g, and mean pore sizes ranging from 3 to 10 μm. Because of the high operating temperatures (~650°C), there is some tendency for sintering of these plaques, particularly on the anode side where the material is generally in a metallic state. Because of cost and stability considerations, nickel- or cobalt-based materials are employed as an anode. Pure nickel or cobalt show significant sintering at 650°C because of a vacancy or surface diffusion, and additives such as Cr or Al_2O_3 ($LiAlO_2$) can be employed to retard the sintering rate. The addition of Al_2O_3 or $LiAlO_2$ is particularly attractive because of its low cost and easy availability compared with Cr.

Copper is a stable material at anode operating conditions, but the stability

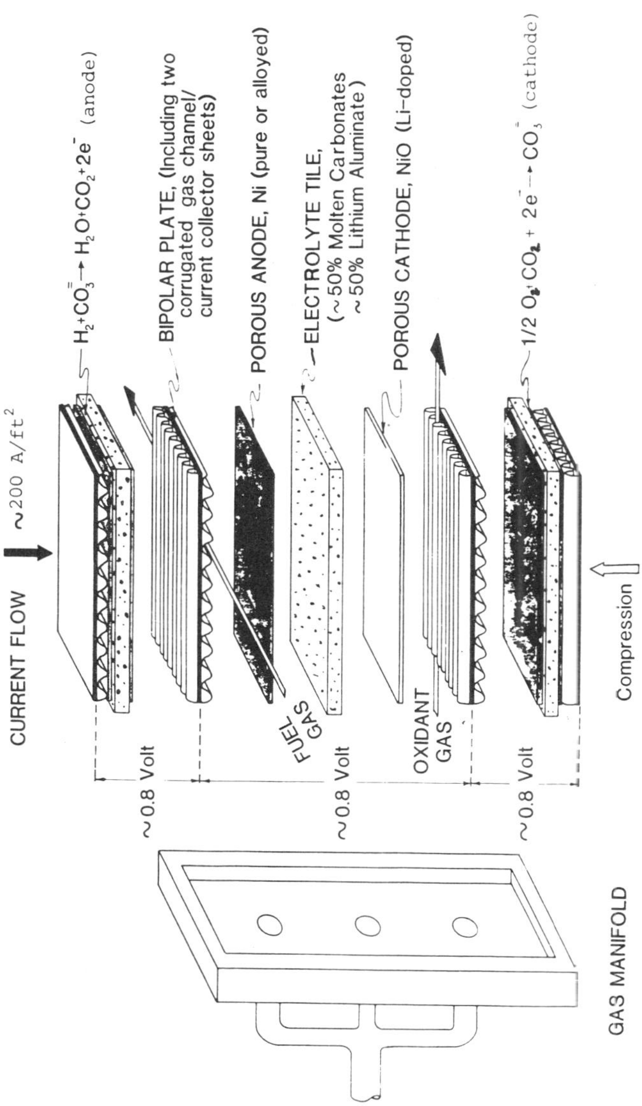

FIGURE 9. An exploded view of a molten carbonate fuel cell stack.[28]

of a porous copper structure would be poor unless some sintering retardant materials are added.

Nickel oxide has been a standard cathode material for the past two decades. Under molten carbonate fuel cell operating conditions, NiO becomes lithiated (3 to 5 atom %) and acquires semiconducting properties. Some solubility and structural instability of NiO have been reported recently.[26] Hence a search for a more stable material is underway.

4.1.3. Electrolyte Support Materials

In a molten carbonate fuel cell (MCFC), the electrolyte matrix is typically a paste formed by the electrolyte melt and a high surface area nonconductive support material. Present day fuel cells use a $LiAlO_2$ powder as the support material. This material exhibits at least three crystal morphologies:

a. α-$LiAlO_2$ (rounded spherical particles),
b. β-$LiAlO_2$ (acicular particles),
c. γ-$LiAlO_2$ (platelets or bipyramids).

Among these, the high temperature γ-$LiAlO_2$ appears to be the most stable phase. Some surface area changes in the material have been observed and attempts have been made to stabilize it by doping with gallium, chromium, etc., but the results have not been too encouraging. Alternative materials are sought, a $SrTiO_3$ material being under study. This material is attractive because it exhibits only one crystallographic phase. Its surface area stability, however, has not been established.

4.1.4. Hardware

Current collector, bipolar plates, and seal materials are required for fuel cell stack construction. The bipolar plate sees an oxidizing environment on the cathode side and a reducing environment on the anode side. Corrosion is particularly severe on the anode side. While Type 316 Stainless Steel is generally satisfactory on the cathode side, with a corrosion rate of approximately 1 mil/yr, the corrosion of Stainless Steel 316 is three to five times greater on the anode side. Nickel or copper plating or cladding has been employed by some workers to overcome this problem. However, a question has been raised about the long term stability of these materials.[27] Because Cr forms a protective layer of $LiCrO_2$ on the anode side, Cr plating or high-Cr alloys may be adequate for this side. Stainless Steel 310 has been found to provide a superior corrosion resistance to Stainless Steel 316.[27] The protective mechanism on the cathode side has been reported to be via formation of $LiFeO_2$. Lithium chromites appear to be soluble under oxidizing conditions and therefore not protective.

A peculiar condition exists at the "wet seal" area of the cell. A wet seal is formed between the electrolyte paste and the bipolar plate edge and it is used to seal the individual gas compartments. Because different gas atmospheres (reducing or oxidizing) and materials (stainless steel, nickel anode, etc.) are present in this area, a variety of corrosion couples are formed,[28] leading to severe corrosion. This problem has been solved satisfactorily by coating the wet seal with aluminum, followed by a treatment to diffuse Al into the bulk Stainless Steel 316. A protective $LiAlO_2$ layer is eventually formed after oxidation and reaction with carbonates (see Equation 3).

Another interesting situation is the gas seal for the manifold supplying gases to a molten carbonate fuel cell stack.[28] This seal should be pliable enough to seal against the edges of the cell stack, it should be nonconductive to prevent any short circuit currents, and it should be stable in a carbonate environment. No single material has been found so far to satisfy these seemingly contradictory requirements. In the absence of a suitable nonconductive gasket material, a bilayer seal consisting of

a. A dense layer of alumina (e.g., "D Gun" coating, Union Carbide) on a stainless steel substrate to provide dielectric properties,
b. A low porosity Al_2O_3 or ZrO_2 felt impregnated with carbonate melt to serve as a pliable seal,

appears to be a practical solution for the time being. Some ionic leakage through the carbonate layer is likely, but a proper design of the felt can minimize this leakage.

4.2. Safety and Storage

Safety considerations include the handling of the salts both at room temperature and in a molten state. Both of these require normal procedures consistent with safe use of chemicals. The alkali carbonates are noncombustible and relatively nontoxic. Li_2CO_3 is used in medicine and Na_2CO_3 and K_2CO_3 are used as general purpose food additives, and therefore their effects on the human body are reasonably well characterized.

A summary of health hazard ratings compiled primarily from the data in Reference 29 is presented in Table VII. As can be seen, the severity of effects increases in the order $Li_2CO_3 < Na_2CO_3 < K_2CO_3$.

The toxic dose for Li_2CO_3 is reported to be 1.7 g/day, and the lethal dose is upwards of 10 g, according to the Lithcoa Material Safety Data Sheet.[30] Similar data for Na_2CO_3 and K_2CO_3 are not available, but the LD50 test (Lethal Dose, 50% fatality) for 47% liquid K_2CO_3 on rats is reported to be 1.87 g/kg.[31] Because the alkali carbonates form a basic solution, there may be a caustic action in the eyes and in the lungs if inhaled. The Material Safety Data Sheet[31] from

TABLE VII. Safety Data for Pure Alkali Carbonates

	Li_2CO_3	Na_2CO_3	K_2CO_3
Melting point, °C	723	858	898
CO_2 equilibrium dissociation Pr.			
mm Hg	10	1	1
@ Temp., °C	730	950	1080
Toxic hazard rating[29]			
Acute local			
Irritant	u	2	3
Ingestion	1	2	3
Inhalation	1	2	u
Acute systemic			
Ingestion	2	u	3
Inhalation	2	u	u
Chronic local			
Irritant	u	1	2
Chronic systemic			
Ingestion	1	u	2
Inhalation	1	u	u
Toxic dose	1.7 g/day (human)	u	Oral LD50 = 1.87 g/kg (47% liquid, rat)
Note	Used in medicine	General purpose food additive	General purpose food additive

Rating: 1 Slight
2 Moderate
3 High
u Unknown

Diamond Shamrock recommends avoiding the simultaneous presence of lime dust (CaO) and K_2CO_3. The combination of these in the presence of moisture or perspiration will cause the formation of irritating caustic potash.

Excessive exposure to the alkali carbonates should be avoided. Where dusty conditions are encountered, a dust mask should be worn. The Lithcoa Data Sheet indicates that for Li_2CO_3, gloves, aprons, etc. are optional but normally not required. Irritation of eyes and skin may be treated by flushing freely with tap water. Medical attention should be obtained if irritation persists.

The other aspect of safe handling concerns the salts in a molten state. An explosive generation of steam due to accidental bulk water carry-over is likely and normal precautions should be taken against such carry-over. When the salts are heated with CO_2 partial pressures that are less than the equilibrium dissociation pressures (listed in Table VII), the salts decompose to form alkali metal oxides:

$$M_2CO_3 \rightleftharpoons M_2O + CO_2 \tag{4}$$

Good ventilation should be provided to prevent irritation of the eyes, skin, and respiratory tract.

The hygroscopic character of the alkali carbonates increases in the order $Li_2CO_3 < Na_2CO_3 < K_2CO_3$. The best possible storage conditions should therefore be provided, keeping the hygroscopicity in mind. Damp floors, leaking roofs, and moist storage conditions must be avoided. Moisture absorption may cause lumping and a reduction in alkalinity. Also avoid storage where contact with acid is possible.

When large quantities are spilled, they should be recovered and may be reused. Small quantities may be disposed of by flushing the area with water into a drain system.

ACKNOWLEDGEMENTS

The careful reviews and suggestions by Dr. B.S. Baker, Professor G.J. Janz, Mr. L. Paetsch, Dr. A. Pigeaud, and Professor J.R. Selman are gratefully acknowledged.

References

1. J.R. Selman and H.C. Maru, "Physical Chemistry and Electrochemistry of Alkali Carbonate Melts," in *Advances in Molten-Salt Chemistry, Vol. 4,* J. Braunstein and G. Mamantov, Eds., Plenum Press, New York (1981), pp. 159–390.
2. J.G. Eberhart, "Molten Carbonate Fuel Cell Wetting Studies," paper presented at First Molten Carbonate Fuel Cell Workshop, Chicago, IL (February 1977).

3. B. Kr. Andersen, "Thermodynamic Properties of Molten Alkali Carbonates," Ph.D. Thesis, The Technical University of Denmark, Lyngby (1975).
4. A.J. Appleby and S. Nicholson, "The Reduction of Oxygen in Molten Lithium Carbonate," *J. Electroanal. Chem. and Interfacial Electrochem.* **53,** 105–109 (1974).
5. Institute of Gas Technology, Fuel Cell Research on Second-Generation Molten-Carbonate Systems, Final Report for DOE Contr. EC-78-C-03-1735, Report No. SAN-1735-4 (April 1979).
6. G.J. Janz, *Molten Salts Handbook,* Academic Press, New York (1967).
7. G.J. Janz, C.B. Allen, N.P. Bansal, R.M. Murphy, and R.P.T. Tomkins, *Physical Property Data Compilations Relevant to Energy Storage, II. Molten Salts: Data on Single and Multicomponent Salt Systems,* National Bureau of Standards, U.S. Department of Commerce, NSRDS-NBS 61, Part II (1979).
8. E.J. Cairns and D.I. MacDonald, "Sensitive Thermal Analysis Establishing Formation of the Incongruently Melting Compound $LiNaCO_3$," *Nature* **194,** 441–442 (1962).
9. M. Rolin and J. Recapet, "Contribution a l'etude des properties thermodynamique de carbonates alcalins, I. Le diagramme ternaire CO_3Na_2-CO_3K_2-CO_3Li_2," *Bull. Soc. Chim. Fr.* **(1964),** 2104–2109.
10. G.J. Janz and M.R. Lorenz, "Molten Carbonate Electrolytes: Physical Properties, Structure, and Mechanism of Electrical Conductance," *J. Electrochem. Soc.* **108,** 1052–1058 (1961).
11. G.J. Janz and F. Saegusa, "Molten Carbonates as Electrolytes: Viscosity and Transport Properties," *J. Electrochem. Soc.* **110,** 452–456 (1963).
12. J. Braunstein and G.D. Robins, "Electrolytic Conductance Measurements and Capacitive Balance," *J. Chem. Educ.* **48,** 52–59 (1971).
13. Institute of Gas Technology, Fuel Cell Research on Second Generation Molten-Carbonate Systems, Vol. II, Final Report for Argonne National Laboratory, Contr. 31-109-38-3552, Chicago, IL (September 1977).
14. U. Merten and W.E. Bell, "The Transpiration Method," in *The Characterization of High Temperature Vapors,* J.L. McGrave, Ed., John Wiley & Sons, New York (1967).
15. M. Schenke, G.H.J. Broers, and J.A.A. Ketelaar, "The Solubility of Oxygen in Molten Carbonates," *J. Electrochem. Soc.* **113,** 404 (1966).
16. A.J. Appleby and C. Van Drunen, "Solubilities of Oxygen and Carbon Monoxide in Carbonate Melts," *J. Electrochem. Soc.* **127,** 1655–1659 (1980).
17. M.A. Volgin, A.L. L'Vov, and V.A. Loskutkin, "Determination of the Diffusion Coefficient of Molecular Hydrogen in Molten Carbonates by Electrochemical Methods," *Sov. Electrochem.* **9,** 353–355 (1973).
18. E.S. Argano and J. Levitan, "A Dual Reference Electrode System for Molten Carbonate Cells with Immobilized Electrolyte," *J. Electrochem. Soc.* **116,** 153–155 (1969).
19. A. Borucka, "Electrochemical Behavior of the CO/CO_2 Gas Electrode in Molten Alkali Carbonates-I. Experimental Technique and Thermodynamic Results," *Electrochim. Acta* **13,** 295–307 (1968).
20. K.Y. Kim and O.F. Devereux, "Use of Alumina as a Sheath for the Pt, $CO/CO_2|Na_2CO_3$ Reference Electrode," *J. Electrochem. Soc.* **125,** 1796–1799 (1978).
21. J. Braunstein, H.R. Bronstein, S. Cantor, D. Heatherly, and C.E. Vallet, Molten Carbonate Fuel Cell Research at ORNL, Report ORNL/TM-5886, Oak Ridge National Laboratory, Oak Ridge, TN (May 1977).
22. Atomics International, "Development of a Molten Carbonate Process for Removal of Sulfur Dioxide from Power Plant Stack Gases," *Progress Report No. 2* to U.S. Public Health Service, Contract No. PH 86-67-128, Canoga Park, CA (1968).
23. H.J. Davis and D.R. Kinnibrugh, "Passivation Phenomena and Potentiostatic Corrosion in Molten Alkali Metal Carbonates," *J. Electrochem. Soc.* **117,** 392–396 (1970).

24. G.J. Janz and A. Conte, "Corrosion of Gold-Palladium, Nickel and Type 347 Stainless Steel in Molten Alkali Carbonates," *Corrosion* **20,** 237t–238t (1964).
25. L.F. Grantham and P.B. Ferry, "Corrosion in Alkali Carbonate Based Melts," in *Proc. Int. Symp. Molten Salts,* J.P. Pemsler, J. Braunstein, Eds., Electrochemical Society, Princeton, NJ (1976), pp. 270–281.
26. H.C. Maru, A. Pigeaud, and L. Paetsch, "Progress in Molten Carbonate Fuel Cell Components Technology," *National Fuel Cell Seminar Abstracts,* U.S. Dept. of Energy, Washington, D.C. (June 1981).
27. K.W. Browall, D.A. Shores, and P. Singh, "Cell Separator Materials for Carbonate Fuel Cells," *National Fuel Cell Seminar Abstracts,* U.S. Dept. of Energy, Washington, D.C. (June 1981).
28. A. Pigeaud, A.J. Skok, P.S. Patel, and H.C. Maru, "Coating Applications for the Molten Carbonate Fuel Cell," *Thin Solid Films* **83,** 449–454 (1981).
29. N.I. Sax, *Dangerous Properties of Industrial Materials,* Third Edition, Van Nostrand, New York (1968).
30. Lithium Corporation of America, *Material Safety Data Sheet,* Gastonia, NC (January 31, 1979).
31. Diamond Shamrock Corporation, *Material Safety Data Sheet,* Plainsville, OH (n.d.).

3

Oxides, Silicates, Phosphates, and Borates

I. D. Sommerville and H. B. Bell

1. Introduction

In view of the fact that the melts discussed in this Chapter have liquidus temperatures mainly in the temperature range 1000–1600°C, it seems appropriate to include a brief section on how such elevated temperatures are obtained and measured and, more particularly, how melts are contained at such temperatures.

Heating is done either by resistance or induction. The resistance can be provided either by a metallic winding which heats by conduction or by nonmetallic rods which heat by radiation. For temperatures up to ~1100°C, Kanthal or Nichrome wire can be used for the winding, but to reach 1500–1600°C platinum or platinum–rhodium alloy windings are commonly used, the proportion of rhodium being increased as the desired temperature is increased. Furnaces wound with Pt–40%Rh can reach 1650°C, while those wound with tungsten or molybdenum can attain even higher temperatures, the limit usually being set by the refractories rather than the winding. The latter are less convenient, since tungsten and molybdenum are easily oxidized at temperatures above 500°C and have to be protected, usually by the passage of cracked ammonia which provides a mixture of hydrogen and nitrogen. The temperature range of wound furnaces can be further extended by using an element which has sufficient strength to be used unsupported. These are normally used in vacuum furnaces in the form of a double spiral. Carbon tube furnaces can be operated at temperatures over 2000°C, but have the disadvantage of being limited by the atmosphere which can be maintained inside them, and in reacting with any oxide which comes into contact with the carbon.

For temperatures of 1400–1600°C, rods are convenient heating elements.

I. D. Sommerville • Department of Metallurgy and Materials Science, University of Toronto, Toronto, Ontario, Canada. *H. B. Bell* • Metallurgy Department, University of Strathclyde, Glasgow, Scotland

Usually made of silicon carbide, and normally four to six in number, these can be connected in different ways to give different total resistances and hence different power inputs. This will almost certainly have to be done as the rods age by conversion of silicon carbide to silica, and their resistances increase. For all types of resistance heating, the temperature is held constant by means of a proportional acting controller.

Induction heating is produced by the passage of current through a coil connected to a high-frequency generator, which induces eddy currents in susceptible material placed inside the coil. The temperature limitation in this method depends on the container used. Where graphite can be used as the container it also acts as the susceptor, the oxides being heated by contact with the crucible. This method is, however, limited to melt compositions and temperatures where there is little reaction with carbon, and the slag is usually crushed and roasted in air to remove any carbon and carbides picked up during melting. In melts which contain easily reducible oxides such as P_2O_5, FeO, MnO, NiO, etc. crucibles of Pt, Ir, Mo, or Fe must be used and in these cases a molybdenum tube can be used as a susceptor. This method is widely used for pre-melting slags to be used in one or another of the experimental techniques to be described later, since it is extremely rapid and a few hundred grams can be melted at a time.

Temperature measurement is either by thermocouples, base metal couples being useful up to 1100°C and precious metal ones up to 1600°C, or by radiation pyrometers of various kinds, which have the advantage of having virtually no upper temperature limit, and which are widely used in conjunction with induction heating where the field set up by the coil may affect the EMF generated by a thermocouple.

With regard to container materials for use at high temperatures, the basic situation is that materials suitable for containing oxide systems are unsuitable for metals and vice versa. Normally slags are held in metal crucibles such as Fe, Mo, Pt, or Ir, while metals are held in refractory oxide crucibles such as SiO_2, Al_2O_3, ZrO_2, CaO, or MgO. This is why slag-metal studies are so much more difficult to conduct than either slag-gas or metal-gas studies.

Metal crucibles can be produced by machining from the solid, welding under neutral atmosphere, or by pressing sheet in a die. Generally SiO_2, Al_2O_3, or ZrO_2 crucibles are purchased, though satisfactory Al_2O_3 crucibles can be made by ramming inside a suitable former. Crucibles of CaO, MgO, and MnO can be produced by slip casting. This involves casting a slurry of fine powder of the particular oxide in either water or alcohol into a split mold. After a solidified skin of the desired thickness has been allowed to form, the rest of the slurry is decanted and the crucible removed by splitting the mold.

In general, to avoid any appreciable reduction of the crucible during experiments, the oxide of the crucible has to be thermodynamically more stable than the oxide of the metal it contains. Also, when oxide melts are held in an oxide container, they become saturated in the oxide of the container, so that the activity of that component is fixed at unity. Obviously, the container must have a melting point above the temperature at which it is desired to conduct the experiments, have reasonable thermal shock resistance (particularly for kinetic experiments), and preferably have low reactivity with the materials to be contained.

The only material which is an excellent container for both slag and metal is graphite, but it has the disadvantage that metal will become carbon saturated and that any reducible oxides in the slag will be reduced. Thus, graphite has been widely used for investigating reactions and thermodynamic properties under reducing conditions, but is of no use for higher oxygen potentials. The way in which this problem has been minimized or avoided will be noted as appropriate in the succeeding discussion.

1.1. Preparation of Melts

The initial step in the study of the phase diagram or any physical or thermodynamic property of a molten silicate, phosphate, or borate is the preparation of samples of accurately known composition. This is usually accomplished by weighing out the required quantities, mixing thoroughly, and melting. As noted already, the choice of container for melting can be a problem since it must be inert to the melt. Preliminary experiments in several systems have shown that pre-melting is not absolutely essential, since the same final equilibrium is achieved whether or not this is done.

Slags containing oxides which are easily hydrated such as CaO or Na_2O are often best made up by weighing the oxide as the carbonate, mixing with other components, and calcining the mixture immediately before melting. Thereafter, all contact with either moisture or carbon dioxide must be rigorously avoided, and where material has been stored for any length of time, recalcination is recommended. Where melts containing P_2O_5 are being prepared, a phosphate should be used wherever possible. For volatile components such as Na_2O, the weight used can be in excess of that required for the final composition in order to allow for some loss during melting, the extent of this being a function of the experimental conditions and only obtainable by experience. In this situation, the preparation and analysis of premelted master slags is essential.

In all cases, the purest components available should be used, which often means the analar grade reagents available commercially, though there are exceptions. The most suitable form of silica for making slags is obtained from natural sources such as white sand or crushed rock. This can be further purified by treatment with dilute hydrochloric acid to remove the last traces of iron. Both FeO and MnO can be obtained in satisfactory purity by decomposition of the respective oxalates at $\sim$1000°C in vacuum or in a CO–CO_2 or H_2 atmosphere, followed by quenching. FeO is also fairly commonly prepared by heating Fe_2O_3 in contact with iron and quenching to avoid reversion. Both FeO and MnO are nonstoichiometric oxides whose compositions will vary with the oxygen pressure, so that melts containing them must be studied under controlled and stated oxygen potentials.

Melts containing oxides of high stability such as the alkali and alkaline earth oxides can be prepared by melting in air, but those containing oxides of the transition metals such as iron, manganese, nickel, etc. require to be made under controlled conditions and ideally under the oxygen pressure of the melt. In the case of ferrous oxide, the system is normally studied at iron saturation. Studies are also made of iron oxide-containing systems at various oxygen pressures, in which case the ferrous and ferric iron contents are reported. These latter systems are noncondensed systems where the gas composition is a phase of importance in the application of the phase rule, whereas systems such as $CaO–SiO_2$ are normally treated as condensed systems.

1.2. Preparation of Gas Mixtures

When studying reaction kinetics or reaction equilibria between gases and condensed phases, the chemical potential of one or more elements in the gas phase can be controlled readily by using appropriate mixtures of gases. The following mixtures have been extensively used:

$H_2–H_2O$, $CO–CO_2$ & $H_2–CO_2$ for controlled oxygen potentials
$H_2–H_2S$ for controlled sulphur potentials
$H_2–H_2O–H_2S$, $CO–CO_2–SO_2$, & $H_2–CO_2–SO_2$ for controlled oxygen and sulfur potentials
$CO–CO_2$ & $H_2–CH_4$ for controlled carbon potentials
$H_2–NH_3$ for controlled nitrogen potentials at $<600°C$.

The calculation of equilibrium partial pressures for these gas mixtures is complicated by partial dissociation of molecular species and formation of other gaseous species which occur at elevated temperatures, e.g., CO, CO_2, COS, S, S_2, SO, SO_2, SO_3, O, and O_2 are all present in a mixture originally composed of CO, CO_2, and SO_2. $H_2–H_2O$ mixtures are sometimes preferred to $CO–CO_2$ mixtures for controlling oxygen potentials, since they avoid the problems of carburization and carbon deposition, but thermal segregation can be more severe.

The required partial pressures are usually obtained simply by mixing gases flowing at different rates, the flowrates being measured on previously calibrated flowmeters. For each component, the partial pressure is equal to the ratio of its flowrate to the total flowrate.

Normally, in precision experiments the gas has to be purified. In purification of O_2, traces of hydrocarbons, CO, and H_2 can be removed effectively by passage over platinized asbestos at 450°C, then through a column of Ascarite (asbestos impregnated with NaOH) to absorb CO_2 and through a column of Anhydrone ($MgClO_4$) to absorb H_2O vapor. The most commonly used drying agents are

SiO_2 gel, Anhydrone, and P_2O_5, the latter being the most effective. O_2 is removed from inert gases and CO by passage over heated active Cu, Ca, or Mg. The main impurities in cylinder CO are moisture and CO_2 which can be removed as already stated. In the absence of O_2, N_2 can be removed from inert gases and H_2 by passage over heated Ca or Mg, thereby forming a nitride.

Gas purification is particularly important where the impurity concerned could poison the surface of the condensed phase and hence change the rate-controlling step of the reaction being studied.

2. *Measurement of the Phase Diagrams of Oxide Systems*

The liquidus temperature of a melt is one of the most important pieces of data about the system, but information on the solid phases involved is also valuable. There are considerable data on oxide systems with silica as one of the components and to a lesser extent on borates and phosphates. Many of the difficulties involved in the study of these systems stem from the high temperatures involved and from the fact that control of the oxygen pressure is often necessary during measurements.

In the case of silicates some workers prefer to prepare gels rather than melts. One of the problems in preparing compositions is the tendency for the melts to form glasses on cooling. This is particularly the case with silicates, and often long annealing times are necessary to give crystalline samples. In studying ternary and higher systems a large number of samples are required, so that an appreciable time is involved in establishing the phase diagram for the whole system.

One of the classical methods of determining liquidus temperatures is by cooling curves or differential cooling curves and this technique has been used for oxide melts. Its use is restricted to melts which do not show significant undercooling and which crystallize on solidification. In some cases "seeding" can be applied to minimize undercooling, but where the melt tends to glass formation cooling curves are not satisfactory.

The quench method has been used for a wide range of oxide melts. This technique involves heating small samples of the crystalline slag to various temperatures and then quenching. The samples are then examined petrologically to assess the degree of melting. This technique can detect both the beginning and end of melting and can also be used to identify the phases involved. This technique can yield accurate data but can be time consuming.

Hot stage microscopes in various forms have been used to measure liquidus temperatures with reasonable accuracy. Welch developed an ingenious design in which a thermocouple is used as the heating element, the support for the melt, and as the thermocouple to measure temperature.[1] A small amount of the sample is placed on the thermocouple tip which must be carefully made to a set profile. The thermocouple tip is viewed through a binocular microscope with a light source projected onto the sample. A current is passed through the thermocouple wire using the half cycle of an ac current; in the other half cycle the temperature is measured (Figure 1). The liquidus temperature is measured by observing the disappearance of the last crystal in the melt. With careful adjustment of the current to the thermocouple a crystal can be made to appear and disappear over a temperature range of a few degrees. The thermocouple is normally contained within a closed cell and an inert atmosphere used. In some cases the primary phase can be identified by microscopic observation. The melt can also be rapidly

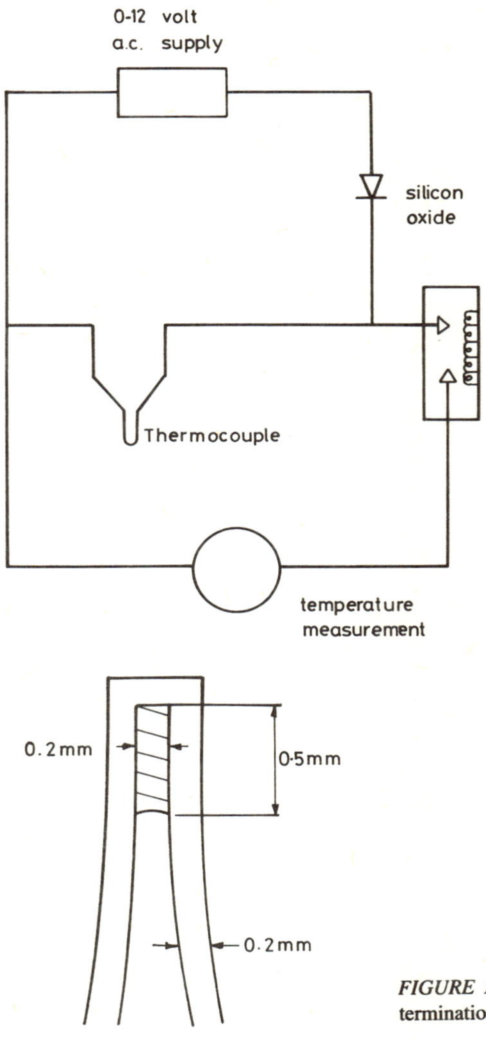

FIGURE 1. High-temperature furnace for liquidus determinations.

quenched by switching off the power and, using X-ray diffraction, the primary phase can be identified. This technique can only be used where the melt is transparent to light and this limits its application to some extent. The technique has been extended by the use of two cells to give a high speed DTA apparatus.[1,2] One cell has the oxide melt while the other has some reference material such as Al_2O_3. Using a programmed cooling rate with a high speed recorder a DTA trace can be obtained. This technique is useful for detecting sub liquidus changes where these are not sluggish. The temperature limitations depend on the thermocouple used and can extend to over 1700°C with platinum–rhodium alloy thermocouples. An iridium hairpin shaped heater has also been used to extend the temperature range.

3. *Structural Determinations in Phosphates and Silicates*

Silicate and phosphate melts are ionic liquids in which the anions can form polymeric ions such as $(P_n\,O_{3n+1})^{-(n+2)}$ and $(Si_n\,O_{3n+1})^{-(2n+2)}$. The distribution of the various ions is of interest in understanding the thermodynamic behavior of liquid slags, and techniques have been developed to study this distribution. These techniques are normally based on a study of the quenched melt and as far as temperature is concerned the measurements are related to the glass-forming temperature. The melt must be rapidly cooled to minimize any changes after glass formation; this is easiest with the more viscous acidic melts.

The procedure for phosphates[3] is a chromatographic one using paper chromatograms. A sample of the quenched phosphate is dissolved in a neutral solution, e.g., distilled water or a 5% aqueous solution of ethylenediamine tetraacetate neutralized by NH_4OH. A small aliquot is then painted on one end of a paper strip and developed at low temperature in a sealed vessel. The developer used is a solution of ammonia, trichloroacetic acid, and acetone in distilled water. After development the strips are dried and sprayed with an ammonium molybdate solution (1 g molybdate, 1 ml HCl, 5 ml 72% $HClO_4$ in 100 ml H_2O). By exposure to ultraviolet light the phosphate is converted to the blue ammonium phospho-molybdate complex and allows separation of the different bands, which are then analyzed for phosphorus. This technique allows separation of up to seven phosphorus atoms in a chain. Above this the separation is not clear enough.

A more complex technique[4] must be used for silicates and the method which has proved most successful in trimethylsilylation. The silicates are leached from the solid and simultaneously trimethylsilylated to give trimethylsilyl (TMS) derivatives. The procedure has been described by Masson *et al.* A solution of 90 ml hexamethyldisiloxane, 10 ml of isopropyl alcohol, and 20 ml trimethylchlorosilane is made up in a closed Teflon jar and stirred. To this solution 1.5 g of the slag are added together with 2 ml distilled water. This mixture is stirred for 5 h and then settled. The organic layer is removed and concentrated by distillation until most of the unreacted trimethylchlorosilane has been removed. The distillation residue is transferred to a Teflon jar and stirred with 1 g Amberlyst 15 ion-exchange resin to complete trimethylsilylation. This requires 12 h or more depending on the slag.

Franks and Inman[4a] have used an electrochemical technique to estimate the proportion of phosphate polymers in sodium metaphosphate melts at 600–810°C. This technique detected the orthophosphate anion, whereas the paper chromatographic technique using quenched samples did not. Franks and Inman conclude that quenching increases the degree of polymerization. This would suggest that the distributions found by chromatography may apply to the glass-forming temperature. There are also problems in the trimethylsilation technique in that changes take place during the treatment of the silicate, e.g., the quantity of water used can affect the results. It is also not clear to which temperature the data refer to in quenched silicates.

A gas chromatograph is used to separate the various TMS derivatives, SiO_4^{4-} at 130°C, $Si_2O_7^{6-}$ at 180°C, and so on, using a temperature rise of about 2°C per minute. Samples of each species are collected and examined in a mass spectrometer to determine the composition of each species. Difficulties arise due to fragmentation of species in the mass spectrometer.

Waseda[5] has studied the X-ray scattering pattern of silicate melts. Using this pattern an electron radial distribution function is calculated and hence the average interionic distances and coordination numbers can be obtained. This technique has not as yet revealed as much quantitative information about ionic species as the chromatographic techniques, but has confirmed the existence of polymeric ions in liquid silicates.

4. *Measurement of Physical Properties of Oxide Systems*

4.1. *Density of Liquid Slags*

There are three main methods used to measure densities in liquid slags. These are the Archimedian method, maximum bubble pressure methods, and the sessile drop method.

The Archimedian method is the normal technique of measuring the upthrust of the liquid on an immersed bob. A material must be used which is inert to the liquid slag and this is one of the main drawbacks to the method, although platinum can be used in many situations.

The maximum bubble pressure[6] method has been used by a number of workers. This is based on the relation

$$\rho_s = \rho_m \left(\frac{H_2 - H_1}{h_1 - h_2} \right)$$

where ρ_s = density of slag, ρ_m = density of manometric liquid, H_2 and H_1 are heights of manometric liquid for immersion depths of h_1 and h_2 respectively.

This method assumes that the surface tension contribution to the bubble pressure is independent of the total bubble pressure. A correction is necessary for the rise in level of the melt due to displacement by the immersed tube. In practice an inclined manometer is used to measure the pressure and a gas must be used which will not react with the melt. This is particularly difficult in the case of melts such as those containing iron oxides where the composition depends on the oxygen partial pressure.

The sessile drop technique is often used to measure the surface tension of liquids but can also be used to measure densities (Figure 2). A substrate must be chosen which will not be "wetted" by the liquid slag. A drop of the slag is melted at a given temperature and a photograph taken of the droplet on the substrate. The drop takes up a shape which depends on: (a) gravity, (b) density, (c) surface energy, and (d) the contact relationship, between the interfacial tensions, slag-gas, substrate-liquid, and substrate-gas.

From measurements on the shape of the droplet the volume can be calculated and by weighing the droplet after measurement the density can be obtained. This technique is often used in conjunction with surface tension measurements. Since it is a technique sensitive to surface properties, the atmosphere is important.

For accurate measurements the Archimedian and bubble pressure methods are to be preferred.

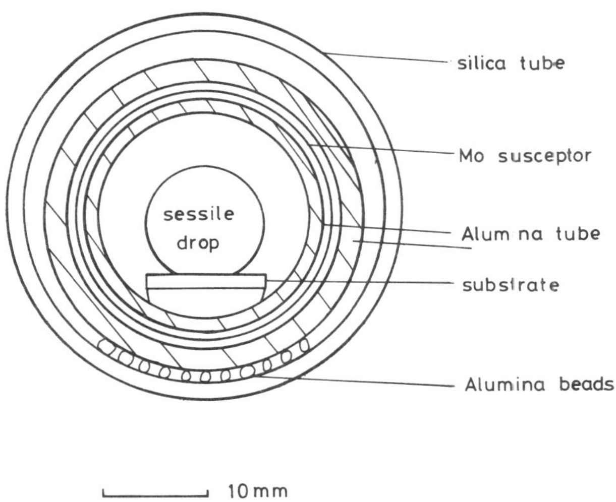

FIGURE 2. Furnace arrangement for sessile drop.

4.2. Viscosity Measurements in Liquid Slags

Viscosity measurements are of value from the point of view of the application of slags and also in an interpretation of the structure. The major difficulties in measuring viscosity stem from the high temperature involved, the corrosive nature of the melts, and the necessity in many cases to control the atmosphere. The preparation of melts has been described earlier.

Viscosity measures the resistance to shear of a liquid and several techniques have been used to measure this property. These include oscillation, rotating cylinder, and falling sphere techniques. The oscillation techniques[7] involve the measurement of the rate of damping of a bob in the liquid slag. This is sometimes called the logarithmic decrement technique. Various methods can be used to follow the damping process, e.g., visual using a galvanometer mirror on a scale, or photographic paper can replace the scale. A light follower may be used to plot the decay rate or electromagnetic measurements may be made. The viscosity is given by the relation

$$\eta = K \frac{lnl}{\sqrt{1 + \left(\frac{lnl}{2\pi}\right)}}$$

$$\text{or } \eta = K\, lnl \text{ where } lnl \text{ is small.}$$

The value of K can be found by calibration using a liquid of known viscosity under identical conditions of immersion depth.

In this technique the rate of shear varies during the measurement. This technique is best for measurements in the range 5×10^{-4} to 5 Nsm^{-2}.

The rotating cylinder technique[8] is somewhat more versatile than the oscillating cylinder method. It is capable of a wider variety of rates of shear and this is constant during the measurement. Normally a bob is hung in the center of the melt and the crucible containing the melt is rotated at a fixed known speed. The shear on the central bob is measured from its angular deflection. The deflection can be read using a mirror and scale or by electrical methods. In some cases the inner cylinder is rotated and the drag on the outer cylinder measured, e.g., modified Brookfield viscometer.

The viscometer is usually calibrated using liquids of known viscosity and corrections made for the effect of temperature on the dimensions of the cylinders. The method is best for liquids with viscosities in the range 10^{-2} to 10^{+3} Nsm^{-2}, but has been used for viscosities lower than this.

In the falling sphere technique[9] the time required for a metal sphere to be dragged through a known height of liquid slag is measured. This is compared against the time taken for the same conditions in air. The method is most suitable for melts with viscosities in the range 3–100 Nsm^{-2}. The technique is an application of Stokes' law—a sphere is attached to one arm of a balance and counterbalanced by weight on the other arm. The sphere is made to rise and fall by altering the weights. The time can be indicated by a pointer on the balance or by using the suspension as the core of a linear variable differential transformer. The viscosity is directly proportional to the time required to move a given distance.

The choice of method depends on the range of viscosities to be measured; the oscillating cylinder method is best for low viscosities. Where one wishes to study the effects of changing the rate of shear, the rotating cylinder technique is the best one.

4.3. *Surface Tension Measurements*

Surface tension of liquid oxides has been measured by three main methods—"ring", sessile drop, and bubble pressure methods.[10]

In the "ring" method a cylinder is usually used rather than a wire ring. The method is based on the pull exerted by a liquid on the cylinder. The cylinder is either allowed to dip into the liquid and then raised until its bottom just reaches the liquid surface or the cylinder is pulled out of the liquid; in each case the pull is measured. The liquid must "wet" the cylinder for this method to be successful.

In the first technique[11] the melt is raised till the cylinder, suspended from a balance, is just immersed in the liquid and then weights put on the balance till the initial position is restored (Fig. 3a). The melt should be in a large enough container so that the liquid surface is level. The surface tension is given by

$$\gamma = \frac{\Delta w g}{2l}$$

where Δw is the difference in weight and l the circumference of the dipping cylinder. The technique is best for melts with low viscosities.

The second technique relies on the cylinder lifting a small amount of the

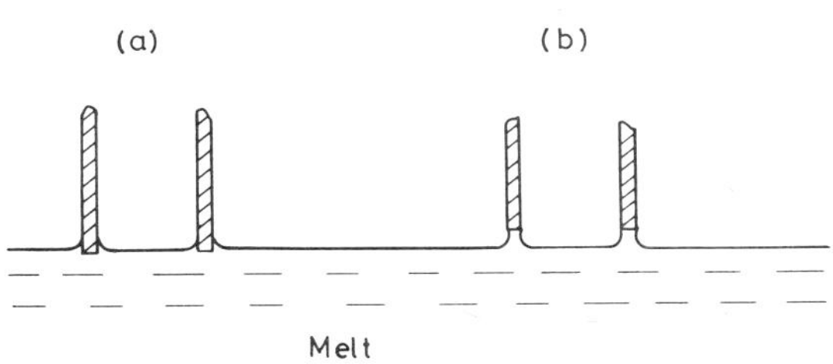

FIGURE 3. Ring methods for surface tension measurement.

liquid and the maximum pull being due to the breaking of the liquid film (Fig. 3b). In this case

$$\gamma = \frac{W_{\max} F \cdot g}{4\pi R}$$

where R is the mean radius of the cylinder and F a correction factor.

The sessile drop method has been referred to under density measurements. A substrate must be used such that the contact angle is above 90°. A number of measurements are made on the drop profile as shown in Fig. 4.

The surface tension is given by

$$\gamma = \frac{\rho \nu^2 g}{\beta}$$

where ρ is the liquid density, ν the radius of curvature at the origin of the coordinate system, i.e., point of maximum height of drop and β is the shape factor which can be found by reference to Bashforth and Adams Tables.

The use of bubble pressure has been referred to under density measurements and its relation to surface tension indicated. When a bubble is blown at the end of a tube immersed in a liquid, its radius of curvature decreases as the bubble expands. If the bubble is small enough to be spherical in form, the maximum bubble pressure is reached when it is hemispherical at the end of the tube. At this point the relation

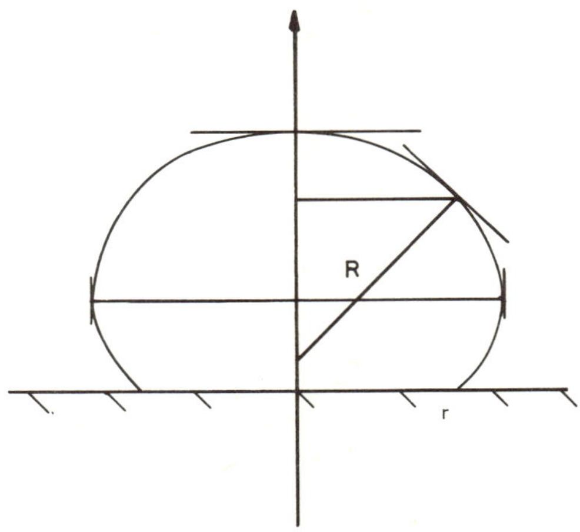

FIGURE 4. Diagram of sessile drop showing measurements

$$P_1 + P_2 = gh\rho + \frac{2\gamma}{r}$$

where P_1 is the hydrostatic pressure and is related to the depth of immersion (h) and density (ρ), and r is the bubble radius. If the contact angle between liquid and the tube is $< 90°$ then r is the internal radius of the tube; if $>90°$ it is the external radius. P_2 is the pressure due to surface tension. The depth of immersion must be accurately known in order to calculate γ, but if two tubes are used with different immersion depths only the distance between the two tubes need be known.

The tubes must be inert to the liquid oxide and must have sharp edges. The bubble must be made to grow slowly to obtain accurate data. The bubble pressure technique has the advantage that a new surface is formed in each measurement whereas in the ring and sensile drop methods the surface remains the same, i.e., any contamination of the surface will affect the measurements.

4.4. Electrical Conductivity Measurements

The most usual method of conductivity measurement is to use a system of electrodes immersed in a melt. Some of the problems involved in this technique are that the resistance measured is low, the electrodes are prone to polarization, and the cell constant is dependent on the immersion depth of the electrodes.

Several arrangements of electrodes have been used and some of these are shown in Fig. 5. The simplest arrangement is to use the container as one of the electrodes as in Fig. 5a, the other electrode being in the center of the cell. While this arrangement can provide electrodes with a large surface area, care must be taken to position the electrode centrally in the cell.

In the two-electrode cell, Fig. 5b, the electrodes are immersed in the liquid at a known distance apart and depth. These electrodes may be wire or strip and the distance between the electrodes should be small compared to the distance of each electrode from the container, to avoid electrical leakage to the container. The disadvantage of this arrangement is the small area of the electrodes which accentuates the effects of polarization.

A variation of the container type is the ring electrode, Fig. 5c. This uses two concentric cylinders as electrodes giving a larger area of electrode and, to a large extent, eliminating stray currents. One drawback is that the areas of the two electrodes are different. In Fig. 5d, the four-electrode arrangement, the two outer electrodes are used to pass a current through the cell and the two inner electrodes to measure the potential. This system has the advantage of virtually eliminating polarization.

An ac bridge must be used to measure the impedance of the cell, and various types have been used. In some designs variable capacitors have been inserted in the bridge, and these increase the sensitivity of the bridge. Some circuits are

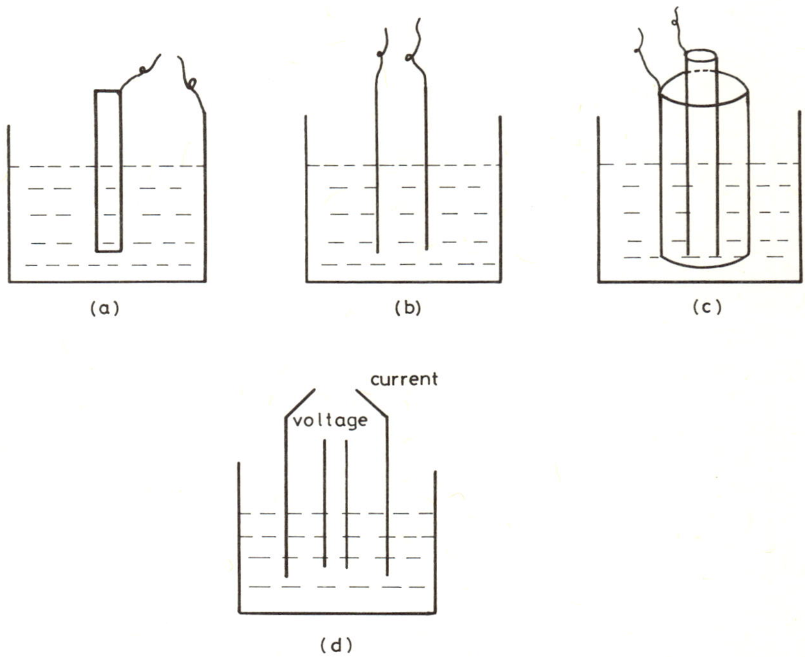

FIGURE 5. Electrode arrangements for electrical conductivity measurements: (a) Central electrode; (b) two-electrode; (c) ring electrode; (d) four-electrode.

shown in Fig. 6. Frequencies of the order of 10 kHz are often used and some workers have shown a variation in the resistance with frequency, the resistance decreasing with increase in frequency, e.g., 30% reduction when frequency changed from 10^4 to 10^2Hz. The capacitance of melts changes by a factor of 1000 in the same frequency range.

Cells are normally calibrated using aqueous solutions of NaCl and KCl. In order to obtain accurate measurements, care must be taken to maintain the distance between the electrodes, the depth of immersion, the position with respect to the container, the liquid height in the container, and the degree of parallelism of the electrodes constant.

A potentiometric system can also be used to measure the cell resistance and this is commonly used with the four-electrode arrangement. The potential and current using an ac source are measured and by applying Ohm's law the resistance is obtained. Small currents must be used in this case to minimize polarization effects.

In measuring the conductivity of molten oxides the design of the measuring system is of great importance as capacitance effects occur, and the conductivity is often a function of the frequency of the measuring current.

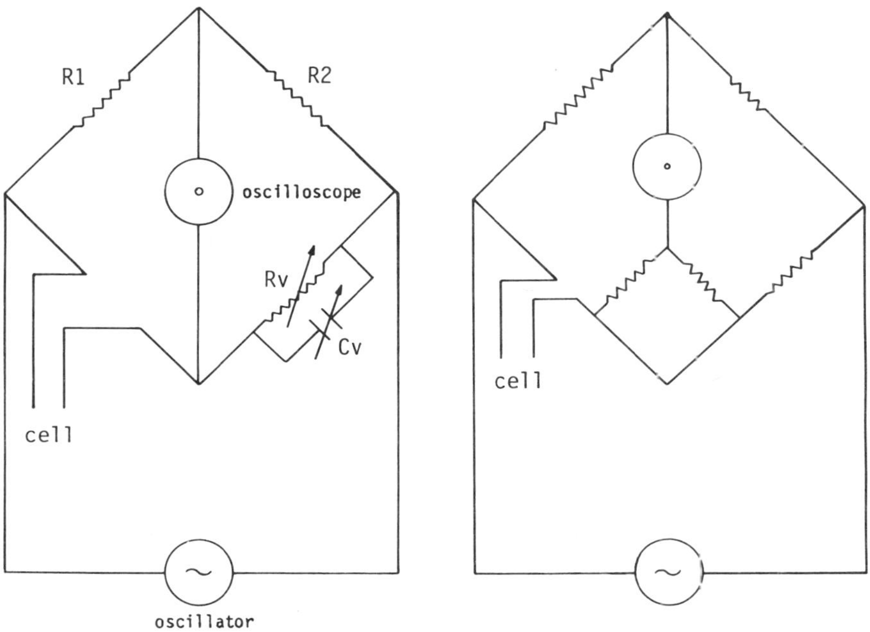

FIGURE 6. Circuits for electrical resistivity measurements.

4.5. Diffusivity Measurements in Slags

Much of the work on diffusivity in molten slags has used radioactive tracers and has been self-diffusivity. The main techniques have been the capillary reservoir method, the instantaneous source method, and the diffusion couple methods.

The diffusion couple method[12] is an adaptation of the technique used for solids. A couple is formed consisting of two liquids of different composition or with liquids of the same composition but one containing a radioactive tracer. The tubes must be of a sufficiently small cross-sectional area to stop convection in the cell. One of the difficulties is that the couples are normally prepared at room temperature and then raised to the diffusion temperature, hence the time for start of diffusion is difficult to fix. This is not normally of importance in solid state diffusion where diffusivities are low (10^{-7} cm^2/s), but it can be of importance in liquids where diffusivities are much higher (10^{-5} cm^2/s). Attempts have been made to avoid this error using a shear cell technique as in Fig. 7. While this technique is attractive there are several difficulties in using it at liquid oxide temperatures. Few containers are satisfactory for liquid slags; graphite and platinum are the most commonly used containers. Graphite could be used for a shear cell but its use is restricted to slags which do not react with carbon at high temperatures. Platinum would be suitable for many slags but at liquid oxide temperatures it tends to weld easily, thus preventing the shear cell being turned.

In the diffusion couple technique the concentration gradient is measured after cooling rapidly from the diffusion temperature. The Boltzmann-Matano interface is located such that

$$D = \tfrac{1}{2}\frac{\mathrm{d}\lambda}{\mathrm{d}c}\int_{C_t}^{C_o}\lambda \mathrm{d}c$$

$$\text{where } \lambda = xt^{-\frac{1}{2}},$$

c is the distance and t the time, C_o and C_t are the concentrations at time zero and time t.

The instantaneous source method has been used for self-diffusivities. In this technique a small amount of a radioactive tracer is placed on the top of a slag contained in a capillary tube; convection must be avoided (Fig. 8). The tube is lowered into the hot zone of a furnace and held there for sufficient time to allow diffusion to take place. The tube is then rapidly cooled. The slag cylinder is then carefully sliced into thin slices and the radioactivity of each slice measured. The diffusivity is found from the following solution to Fick's Law

$$C = \frac{Q}{\sqrt{\pi D t}}\exp\frac{-x^2}{4Dt}$$

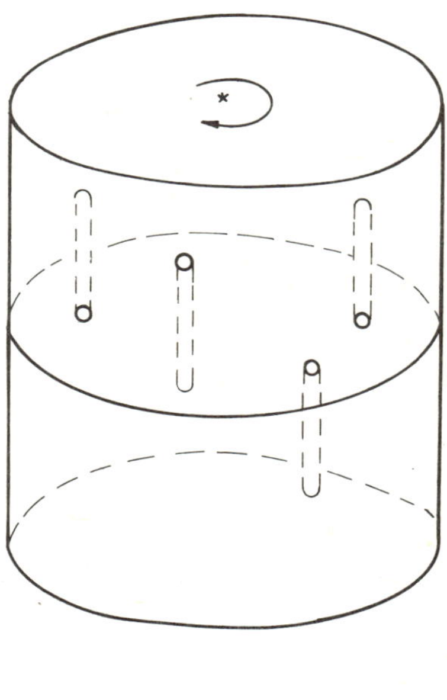

FIGURE 7. Shear cell for diffusion measurements.

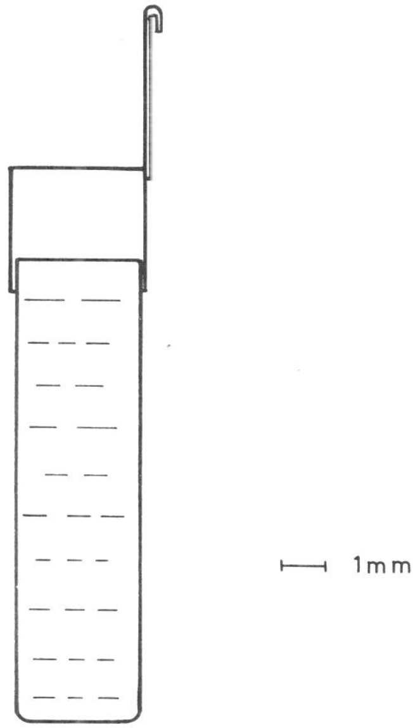

FIGURE 8. Capillary used for instantaneous source technique.

where C is the concentration of radioactive species at distance x, Q is the number of radioactive atoms deposited, t is the diffusion time, and D the radioactive diffusion coefficient. This technique can also be used with a nonradioactive species, but the difficulty is accurate analysis of the concentration profile. There are a number of problems associated with this technique involving decay times of the radioactive species and geometry of the counting method chosen.

The capillary reservoir technique[13] has been used in a number of cases, again mainly for self-diffusivity measurements, although it can be used for interdiffusivities. This technique involves filling a capillary with liquid slag from a reservoir. This is normally done by evacuating the system, lowering the capillary into the slag, and then slowly raising the pressure to force the liquid into the capillary. This capillary is then transferred to a liquid reservoir filled with a liquid of different composition or labelled with a tracer. After a given period of time the tube is removed from the reservoir. Normally the total amount of tracer in the capillary is measured rather than sectioning and obtaining a profile.

The exact solution to the diffusion equation for these conditions is complex but for the situation where diffusion is of a radioactive species into a capillary the approximate solution is

$$D = \frac{1.2732}{td^4\rho^2}\left(\frac{I_i}{S_i^o}\right)^2$$

where I_i is the counting rate of the radioactive species in the capillary and S^o

the specific counting rate in the reservoir, t is the time in seconds, ρ the density of the liquid, and d the diameter of the liquid column in cm.

It is possible by using different species to determine the diffusivity of two species in a single experiment, e.g., Na and P in sodium phosphates.[13] In order to keep the boundary conditions constant at the mouth of the capillary the reservoir must be stirred or the capillaries rotated, the latter method being the more common.

The capillary reservoir technique is best used for relatively fluid melts and for melts where metal containers can be used. For tracer diffusivities the simplest method experimentally is the instantaneous source method.

Permeability measurements of oxygen in liquid slags have been carried out using a thin film of slag adhering to the end of a tube.[14] Oxygen is passed over one side of the film and permeates into an argon gas flow on the other side (Fig. 9). The oxygen content of the argon is measured using an oxygen concentration cell based on a solid electrolyte (ZrO_2 + CaO or ThO_2 + Y_2O_3). The profile of the slag film must be measured to calculate the permeability. A tube must be used which will be wetted by the slag in order to maintain a film.

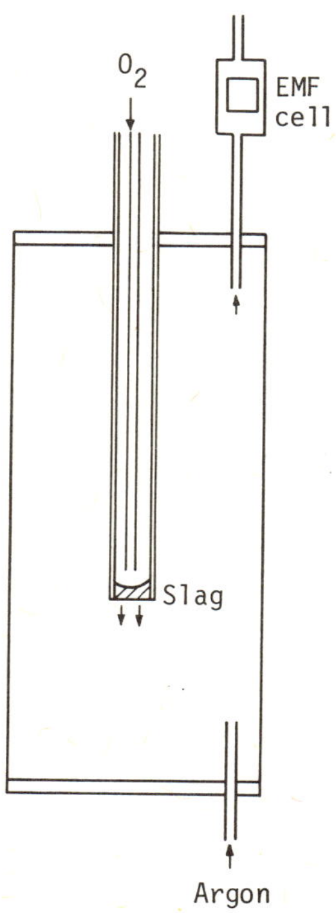

FIGURE 9. Permeability of oxygen in liquid slags.

5. Measurement of the Thermodynamic Properties of Oxide Systems

Nearly all experimental studies of the thermodynamics of silicate systems have been concerned with the determination of activities or partial molar free energy values for one component. Calorimetrically determined partial enthalpies of solution or mixing are available for only a very few systems so that calculation of entropies of mixing or correlation of data for the solid and liquid are not generally possible.

The enthalpies of formation of many silicates have been determined by solution calorimetry. More recent work has used molten oxides and is in general more reliable. Standard entropy values have normally been obtained from low-temperature heat capacity studies while enthalpy values measured by drop calorimetry have provided heat capacities in the temperature range 300–1800 K.

As far as borates and phosphates are concerned, data on partial molar properties have been determined and published for only a very few of the simplest systems. In all three cases, silicates, borates, and phosphates, it is the binary system with PbO which is the most thoroughly documented thermodynamically.

In view of this, the experimental methods described are those which have been used for silicates, although most could be applied to borates and phosphates with suitable modifications. For example, since phosphates are extremely hygroscopic, all traces of moisture must be removed from the system and obviously gas mixtures involving water vapor are not appropriate. In general, the borates and phosphates melt at lower temperatures so that they could be easier to deal with experimentally, but they are of considerably less significance as far as metallurgical industry is concerned and so have not received much attention.

Spencer[15] has reviewed the thermodynamic data available on silicates up to 1973, and this has been brought up to date in a recent publication by the authors.[16] Kubaschewski[17] has presented selected thermodynamic data on a number of double oxide systems, excluding borates, while Slough and Jones[18] have reviewed the sparse thermodynamic data existing on borate systems up to 1974.

It is generally accepted that experimental difficulties increase rapidly, perhaps even exponentially, with temperature, so that appreciably wider error bands are associated with the values obtained in the melts to be discussed in this chapter than is the case for lower melting salt mixtures.

5.1. Measurement of Equilibria between Gas and Molten Slags

Such measurements can be made in either flowing or closed systems. In the case of a flowing system, measurement of change of sample weight is not normally suitable for application to melts involving several components and the normal method is to measure change of slag composition. Where a closed system is used, measurement of change of gas composition in a recirculating system, although possible, has not been extensively used, the more normal methods being measurement of the change of pressure of a single component gas or some technique involving vapor transport. Although considered to be gas–slag equilibrium techniques, some of those methods involve the solid metal phase also, as either reactant or solvent.

5.1.1. Experimental Techniques

Experiments of this nature can be conducted in either horizontal or vertical furnaces, usually with a hot zone of 6–8 cm, the temperature being measured and controlled by precious metal thermocouples. The working tube is usually either recrystallized alumina or mullite, and water-cooled at the ends. The horizontal arrangement (Fig. 10) has the advantage of allowing greater productivity, since it is easier to equilibrate slags in groups of three or four at a time, the separate crucibles sitting in indentations in a fired alumina tray. This has to be introduced either at low temperatures or very slowly at higher temperatures to avoid thermal shock of the furnace tube.

In either case, while the furnace is heating to the desired experimental temperature, argon or nitrogen gas is passed slowly through it and sufficient time is allowed to ensure that the slags are molten before the active gases are introduced. Preliminary experiments are conducted to establish the required equi-

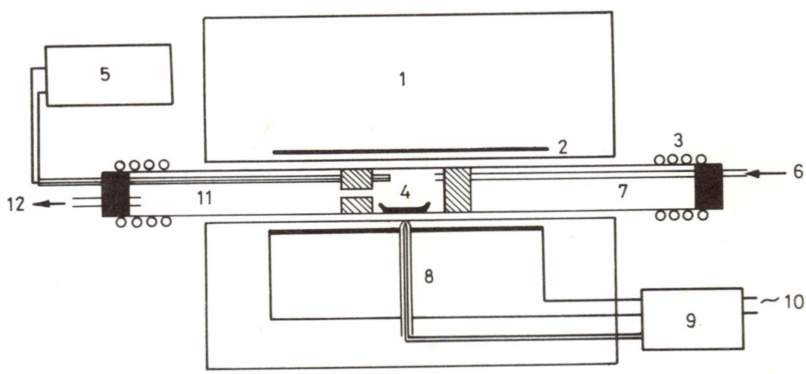

1 - Furnace body
2 - Heating elements
3 - Cooling coils
4 - Sample tray
5 - Potentiometer
6 - Gas inlet
7 - Reaction tube
8 - Controller thermocouple
9 - Temperature controller
10 - Power input
11 - Mullite sheath and thermocouple

FIGURE 10. Furnace arrangement for gas-equilibration work.

libration time, and the time used thereafter is well in excess of this minimum. Once equilibrium has been attained the active gases are turned off, and the slag samples cooled. In order to avoid any change of composition, this cooling has to be as rapid as possible and is normally accomplished by rapid withdrawal of the slag samples into the cool end of the tube or even their complete removal from the furnace. Although this latter method involves a large increase in oxygen potential, the temperature drop is sufficiently rapid to ensure that kinetic factors inhibit any appreciable oxidation of the samples. After cooling to room temperature, the slags are removed from the crucibles, crushed to a suitable particle size and stored in sealed containers ready for analysis.

5.1.2. *Equilibration under Controlled Oxygen Potential*

(a) Equilibration with Container. In this case the crucibles are normally made of iron, such as Armco Iron which is free of any significant amounts of impurities, and the reaction occurring is eminently simple:

$$[\mathrm{Fe}] + \tfrac{1}{2}\mathrm{O}_{2(g)} = (\mathrm{FeO})$$

where [ ] indicates the metal phase and ( ) the slag phase, and the pO_2 is generated by one of the mixtures mentioned previously. The unit activity of iron and the known pO_2 together fix the value of a_{FeO}, and analysis of the slag allows this value to be associated with a particular composition.

This technique is limited to ~1475°C, the maximum advisable temperature for prolonged holding in iron crucibles. Since equilibration is with the sole component of the crucibles, no diffusion in the solid state is required, and experience has shown that 3 h is normally ample time for equilibrium to be achieved. To ensure this, the mixtures are made up reasonably close to the final expected equilibrium compositions, which, after a few experiments have been completed, can be deduced from previous results at similar compositions. The final FeO content is always approached from lower rather than from higher values, i.e., by oxidation of iron rather than by reduction of FeO, since the small iron globules which are produced by the latter approach remain suspended in the slag and are difficult to separate efficiently, even by magnetic means. Very few unremoved globules are sufficient to cause a large error in the analyzed FeO content of the slag, since the analysis is for total iron and does not distinguish metal from oxide.

(b) Equilibration with Foil. Using platinum crucibles to hold slags containing MnO under a controlled oxygen potential allows the reaction

$$[\mathrm{Mn}]_{\mathrm{Pt}} + \tfrac{1}{2}\mathrm{O}_{2(g)} = (\mathrm{MnO})$$

to occur, and temperatures up to 1600°C can be employed. Equilibrium is normally approached by reduction of MnO so that manganese dissolves in the platinum. Since manganese can only enter the crucible wall from one side, and the crucible may well be required for further experiments, a piece of thin platinum foil is usually inserted into the slag so that diffusion can occur from both sides and equilibrium be achieved more quickly. Even so, because solid state diffusion is involved, 6 h is normally allowed. After equilibration, analysis of the slag for MnO and of the platinum for manganese yields the values of γ_{MnO} at each composition used.

In this instance the determination of the thermodynamics of the slag system obviously depends on a prior study of the activity of manganese in the Pt–Mn system, about which there is some disagreement in the literature, such that this is one possible source of error. The main experimental source of error is in separation of the platinum foil from the slag; the foil has, in fact, to be very carefully cleaned using emery paper, of any adhering slag since any slag left attached will cause a significant error in determining the manganese content of the platinum.

This method has also been used for the determination of FeO activities at temperatures of 1500°C and above, although here again there is some doubt about the solution thermodynamics of the Pt–Fe system. In principle, this method can be employed for the determination of the activity of any oxide M_xO_y in a slag which yields a sufficiently high content of M dissolved in platinum at attainable oxygen potentials to make accurate analysis possible, provided that the activity–composition relationship in the Pt–M system has been determined. The other obvious limitation where the crucible is of platinum is that Pt–M alloy obtained at equilibrium should be solid at the experimental temperature, so that the relevant phase diagram should be used in conjunction with the thermodynamic data for the alloys to determine the range of activities in the slag phase obtainable by this technique. It is for this reason that iridium is sometimes preferred to platinum as the crucible material, particularly at low oxygen potentials where significant reduction of silica is expected to occur.

5.1.3. *Equilibration under Controlled O_2 and S_2 Pressures*

The sulfide capacity of a slag is denoted by C_s and defined by the expression

$$C_s = (\text{wt \% S}).\left(\frac{p_{O2}}{p_{S2}}\right)^{1/2} = \frac{(\text{wt \% S})}{A}$$

where $A = \left(\frac{p_{S2}}{p_{O2}}\right)^{1/2}$ and is known as the sulfurizing potential. Thus the evaluation of sulfide capacities requires that both the O_2 and S_2 potentials be controlled simultaneously by use of one of the mixtures mentioned. In general, where the sulfide capacity of the slag is expected to be low, a high A value is used to ensure that there is a sufficiently high S content in the slag to permit accurate analysis, and this is achieved by the use of H_2–H_2O–H_2S mixtures. Where the sulfide capacity is reasonably high, lower A values achieved by using CO–CO_2–SO_2 or H_2–CO_2–SO_2 mixtures will lead to an adequate S content in the slag. Another limitation is placed on the gas mixture used by the fact that the pO_2 generated must be low enough that all the S enters the slag as S^{2-} ions rather than as SO_4^{2-} ions. This limiting value varies with temperature, from $\sim 10^{-6}$ at 1500°C to $\sim 10^{-10}$ at 1300°C. Thus, before experiments can commence, preliminary calculations must be carried out to ensure that the mixture used complies with these restrictions. The calculation of p_{O_2} and p_{S_2} generated by any given mixture is done by a process of successive approximations, so that it can be greatly speeded up by the use of a simple computer program.

The crucibles used are either platinum or iridium, so that temperatures up to 1600°C can be used and equilibration times of 3–4 h are normally adequate. After rapid cooling and crushing, analysis of the slag for sulfur completes the data necessary to calculate its sulfide capacity, which is then linked to its determined composition. The lower temperature limit of this technique is set at around 1200°C by the fact that at lower temperatures the various components of the gas mixture do not equilibrate among themselves in the short time available before they pass over the slag sample, so that the p_{O2} and p_{S2} values in contact with the slag are unknown.

If it is assumed that the presence of a small quantity of sulfur in solution in the slag does not materially affect the activities of components such as FeO or MnO, the insertion of a piece of platinum foil permits evaluation of their activities at the same time as the sulfide capacity, provided that a longer equilibration time is allowed. CaO activities can also be obtained from the sulfur solubility, as discussed later.

5.1.4. Experimental Problems

In addition to the problems and sources of error already mentioned briefly, there are two major problems associated with these techniques. The first is thermal diffusion within the gas mixture, which is the tendency of components of different molecular weight to segregate when passing through a temperature gradient, so that the gas composition presented to the samples is not the equilibrium one. Obviously this problem is greatest for mixtures containing H_2, but it can be overcome in either or both of two ways. One way is to lead the gases through narrow tubes to the reaction zone so that the linear flowrate is increased for a given volume flowrate, although the gases must still be allowed to attain the equilibrium temperature. The other method is to increase the volume and hence the linear flowrate by addition of argon, 50% Ar being a common figure, which also has the effect of increasing the average molecular weight of the mixture. Since it is so important to exclude any thermal diffusion, the two methods are often combined.

The second problem is the tendency of some slag compositions to creep partially or even totally out of the containing crucibles. Unfortunately, this phenomenon is not at all well understood and no complete solution has been produced thus far. Varying degrees of success in inhibiting it have been achieved by placing spiral coils of wire or tents, usually of Pt, in the slag so that they protrude above the surface and guide the creeping slag back into the crucible, or by some form of hat or lid on the crucible, but there are compositions for which none of these measures is effective. The gas plays some part in the matter, presumably by influencing the surface tension of the slag, since it has been observed to be worse for lower O_2 potential mixtures and in particular for those containing H_2. Because of our lack of understanding of the mechanism, it is virtually impossible to predict how serious it will be in any given slag system, so that it has to be discovered and dealt with by experimental methods.

5.1.5. Pressure Techniques

The activities of both SiO_2 and CaO have been determined in various oxide and oxide–fluoride systems by measuring the equilibrium pressures of CO generated by their reactions with carbon

$$SiO_2 + 3C = SiC + 2CO$$

$$\text{and } CaO + 3C = CaC_2 + CO$$

Lime activities can, of course, only be obtained in systems not containing SiO_2.

The method involves holding a finely ground mixture of the slag and carbon in a graphite crucible at the desired temperature until equilibrium is attained in a closed system, the pressure being measured on a manometer. Thus, while it is simple in principle, considerable care and expertize are required to achieve accurate results.

A vacuum-tight system is absolutely essential since even the smallest leak can lead to considerable error in the measured pressures. If the tube or thermocouple sheath are of mullite, contact with the graphite crucible must be avoided or the mullite will quickly become permeable, so that the graphite crucible is normally supported on an alumina plaque or placed inside an alumina crucible. Before each equilibration, the crucible, plaque, and tube all have to be thoroughly degassed to remove all occluded gases. To facilitate this, the crucible should be as thin walled as possible, and charging and loading into the furnace should be done as rapidly as possible to reduce contact with air to a minimum. If either gaseous SiO or Ca vapor are allowed to diffuse away from the reaction zone and condense in a cooler part of the system, true equilibrium will not be obtained. To minimize this, the graphite crucible has a tightly fitting lid through which there is only a fine capillary. This allows rapid equalization of pressure between the crucible and the rest of the system, but slows down the diffusion of these species sufficiently for true equilibrium to be reached. In principle, it is possible to check for all such effects by approaching equilibrium from both sides and so bracketing it accurately; an approach from lower pressures is easier experimentally, since the rate of gas evolution is higher than that for gas absorption. Small random fluctuations in pressure can be a problem, particularly when the reaction rate and therefore the pressure change are very slow. They can make it difficult to decide when equilibrium has been reached, but repeated determinations will yield an accurate result. The addition to the apparatus of buffer volumes which can be introduced or excluded simply by turning stopcocks will increase the range of pressure changes and hence the range of temperatures which can be handled, particularly for those reactions with large temperature coefficients.

The activities of PbO in $PbO–SiO_2$ melts have been measured by holding samples of slag and liquid lead in separate crucibles at the same temperature until both are in equilibrium with the same pressure of PbO vapor. This is an example of the isopiestic technique, which is more normally applied to metals and alloys held at different positions in a temperature gradient.

Since the activities are obtained as the ratio of the oxygen content of the lead in equilibrium with the silicate to that in equilibrium with pure PbO, it can also be regarded as a variant of the methods for measuring oxygen distribution between slag and metal. It also represents one possible way of dealing with the container problem for systems containing a component with a sufficiently high vapor pressure at the experimental temperature, such as the oxides of lead or tin.

When a flowing system is employed, the technique combines features of the transportation and isopiestic methods. An example of this is the method used by Richards and Thorne[19] to determine the activities of both FeO and ZnO in silicate melts by holding them in iron crucibles at 1200°C until equilibrium is achieved with gas mixtures containing known partial pressures of zinc and oxygen. The required p_{Zn} of about 0.03 atm was obtained by passing the known mixtures of CO, CO_2, and N_2 through a bed of zinc oxide pellets at 900°C. One of the problems with this technique is that the amount of zinc in the gas is limited and the resulting ZnO content of slags is around 1%, as compared with over 5% ZnO in industrial slags. It is also a complex way in which to determine the gas composition, and can be subject to errors.

Methods of determining vapor pressures by measuring rates of evaporation have not been greatly used in oxide systems, mainly because of the complexity of the vapor produced and the consequent difficulty of determining the molecular weight of the effusing species. The Knudsen cell-mass spectrometer combination is in principle able to deal with this situation, but because of the very considerable problems both in experimentation and in interpretation of the results, very few successful studies have been carried out. One of these is the study of molten sodium–potassium silicates by Belton *et al.*[20], where it was established that both the sodium and potassium oxides vaporize predominantly by dissociation and that the measured ion–current ratios were representative of the gas–liquid equilibrium with respect to the pressures of Na^+ and K^+. The Knudsen cells were constructed either of dense graphite or molybdenum, with orifice areas of from 1.3 to 7.0×10^{-4} cm^2.

5.2. Measurement of Equilibria between Metal and Molten Slags

Experiments involving two condensed phases are almost invariably carried out in vertical furnaces since this allows the use of larger crucibles, and hence larger masses and interfacial areas of reacting phases. The use of larger masses is particularly important where samples are to be taken during the course of equilibration, so that the proportion removed can be kept low, while interfacial areas should be maximized to reduce equilibration times. Also, for equilibrium to be attained in a reasonable time, forced convection is usually necessary and can be provided either by mechanical stirring or by gas bubbling, provided that there is a sufficient volume of one or both phases present. Preliminary experiments are often required to ascertain these equilibration times and whenever possible equilibrium should be approached from both sides to ensure that no complicating side or coupled reactions are preventing true equilibrium from being reached.

When, as is usually the case, chemical analyses of both phases are required, rapid cooling is essential to minimize any change of composition. Where small quantities are being used cooling of the whole system can be made sufficiently rapid by withdrawing the crucible into the cooled end of the furnace tube. This is satisfactory where the only information wanted is the final equilibrium. If the reaction rate is also of interest, larger quantities are used and sampling is carried out at the experimental temperature. For the slag phase, this can be accomplished by dipping iron or copper rods into it so that a solidified layer adheres and can be withdrawn, while metal samples are obtained by suction into a silica tube. Where necessary, precautions have to be taken to avoid contamination of the slag by the silica of the tube, and allowance has to be made for the amounts of material removed when calculating the reaction rate constant from the analyses obtained.

Most of this work is carried out with both phases liquid. However, the technique of equilibration with solid foil has been extended to alloy foils, an example being the determination of the free energies of mixing in iron–cobalt orthosilicate melts by equilibration with iron–cobalt foils at 1450°C.[21] Up to four alloy compositions coiled into spirals could be equilibrated at the same time so that equilibrium was approached from both sides. For this technique some previous idea of the interdiffusivities in the alloys and their variation with composition is essential so that adequate equilibration times can be used for each alloy composition.

When both phases are liquid the container problem is at its most acute, and graphite is the only really satisfactory material. This problem has been circumvented by the use of a rotating crucible or the levitation technique. The advantages and limitations of this latter method, and particularly the influence of thermal diffusion, have been discussed by McLean[22] for the gas–metal situation, and very similar considerations apply when slags are involved.

By far the most important technique is the classic one of distribution of an element between the two liquid phases until at equilibrium its activity in each is the same or until the activities of the element and its oxide are linked by the known free energy of formation of the oxide. As an example of this phenomenon, the distribution of silicon between slag and carbon-saturated iron has been widely used in the determination of silica activities in slags, the activity of silicon in iron in the relevant concentration range having been previously measured by equilibration with silver–silicon alloys.

In some cases it is the value of the distribution coefficient which is of interest, as for sulfur and phosphorus. Most of the work on sulfur distribution has been done under reducing conditions in graphite crucibles under a CO pressure of one atmosphere. Under such conditions reducible oxides such as FeO, MnO, and even SiO_2 can be reduced as the sulfur is transferred. The effect of dissolved manganese and silicon on f_s† can be bypassed by simply defining the distribution coefficient, L_s, as $\frac{(\text{wt }\%\text{S})}{[\text{wt }\%\text{S}]}$ where, as before, ( ) represents the slag phase, [ ] the metal phase. However, if the transfer of sulfur becomes coupled to SiO_2 reduction, the time required for final equilibrium to be reached will greatly increase because SiO_2 reduction is so slow. Phosphorus distribution, on the other hand, requires oxidizing conditions to retain some phosphorus in the slag phase, and CaO or MgO crucibles are normally employed.

Obviously, the distribution of oxygen also requires higher oxygen potentials and has been used frequently to obtain the activities of FeO in slags from the expression

$$a_{\text{FeO}} = \frac{[\text{wt }\%\text{O}]}{[\text{wt }\%\text{O}]_{\text{sat}}}$$

where $[\text{wt }\%\text{O}]_{\text{sat}}$ is the oxygen content of metal in equilibrium with pure FeO at the same temperature. The use of such an expression, which ignores the

†f_s = activity coefficient of S in liquid iron, referred to a standard state of an infinitely dilute solution of S in iron.

activity coefficient of oxygen, is only valid where the metal, usually iron, is free of other solutes which interact significantly with oxygen. The concurrent distribution of manganese and oxygen allows the calculation of MnO activities also. For very low FeO activities, where the oxygen content of the metal would be too small for analytical accuracy, the values of a_{FeO} can be obtained by combining sulfide capacity and sulfur distribution data.[23]

5.3. Measurement of Solubilities in Molten Slags

The solubility of a gas in ionic melts may occur in either a "chemical" or a "physical" form. Chemical solubility involves chemical reaction of the dissolving gas to form ions in the molten phase, and is displayed by gases such as S_2, H_2O, and CO_2. Physical solubility is due to the accommodation of gas atoms or molecules in voids or "holes" in the liquid structure, and applies to gases such as He, Ar, and N_2. The values are in general appreciably smaller than for chemical solubility, and the two can be distinguished by investigating the dependence of the solubility on melt composition and temperature. Oxygen appears to exhibit chemical solubility in $MO–SiO_2$ melts where M can adopt more than one valency, and physical solubility where M exists in only one oxidation state.

The solubility of sulfur is obtained by a slight extension of the technique previously described for studying sulfide capacities, the sulfurising potential of the gas mixture being increased until saturation of the melt with sulfur is obtained, when there will be a break in the plot of (%S) vs. $\left(\frac{p_{S2}}{p_{O2}}\right)^{1/2}$. This is best done by approaching the equilibrium sulfur contents from both above and below until a pair of gas compositions are found, one of which produces a large increase and the other a large decrease in the slag sulfur content, so accurately bracketing the saturation value.[24] Alternatively, the solubility limit can be obtained by graphical methods.[25] In melts containing CaO, $a_{CaS} = 1$ at this point and the activity of CaO can be obtained from the gas composition and the free energy for the reaction (CaO) + $\frac{1}{2}S_{2(g)}$ = (CaS) + $\frac{1}{2}O_{2(g)}$

In most studies of the solubility of water, slags have been exposed to gas mixtures with a controlled water vapor pressure, sampled, and subsequently analyzed for hydrogen by a vacuum fusion technique. The value of p_{H_2O} is determined by passage through a water bath of an accurately controlled temperature, condensation being avoided by passing the gases through copper tubes maintained at 100°C by heating tape or some similar device. On the assumption that OH^- ions are formed in solution in the slag, the solubility of water is obtained by multiplying the analyzed hydrogen content by nine, a procedure which has resulted in considerable scatter within the experimental data available. Zuliani *et al.*[26] have recently described an elegant thermogravimetric technique which allows precise measurements of both the kinetics of adsorption and desorption as well as the water solubility in liquid slags. Since the weight change is measured continuously on an electrobalance the need for sampling is eliminated and much greater precision is obtained. Another continuous technique has been developed by Stermsek and Lange,[27] whereby the change of water vapor concentration in the melt during absorption and desorption is followed by measuring the pressure change in a closed chamber under isothermal and quasi-isobaric conditions.

The solubility of oxygen in $PbO–SiO_2$ slags, and its dependence on oxygen pressure, slag composition, and temperature have been studied by a volumetric method. The experimental arrangement included two manometers, one to set the pressure level and one to measure the decrease in pressure as oxygen is absorbed. The volume absorbed is obtained by noting the volume required to restore the original pressure.

5.4. *Measurement of Basicity of Molten Slags*

Fairly recently, Wagner[28] suggested that a more fundamental measurement of basicity should be sought in order to correlate equilibrium data for different reactions involving O^{2-} ions. Since, as noted earlier, CO_2 solubilities have not been determined for any large number of melts, his suggestion of using carbonate capacity cannot be widely applied. An alternative approach, developed by Duffy and Ingram[29] for glasses, is to measure spectroscopically the "nephelauxetic" effect, which is the lowering of the frequency of the absorption band for the ${}^1S_o \rightarrow {}^3P_1$ transition of probe ions such as Tl^+, Pb^{2+}, and Bi^{3+} when basicity is increased, due to electron donation by oxygen ions. The spectroscopic shift represents the donor power of the oxides in the glass and the ratio of this to the donor power of free oxide ions has been termed the optical basicity, Λ.

For determination of optical basicity by spectroscopic examination, the fragments of glass need only be a few mm in cross-section since they are mounted over a 3–4 mm dia. aperture of a mask placed in the light beam of the spectrophotometer. The path length of the glass should be ~1 mm to keep absorbance of the glass as low as possible. According to Duffy and Ingram, the frequency of the absorption maximum can usually be determined to within $\pm$ 50 cm^{-1}, corresponding to $\pm$ 0.001 unit. For carrying out determinations on a number of glasses the "platinum loop" technique is very useful.

From these measurements basicity moderating parameters, γ, are assigned to different cations with γ_{Ca} chosen as 1.0 and the optical basicity calculated as

$$\Lambda = \frac{X_A}{\gamma_A} + \frac{X_B}{\gamma_B} + \ldots$$

where X_A and X_B are the equivalent cation fractions of A and B. The application of this approach can conveniently be extended since values of γ can be calculated from the Pauling electronegativities, X, by the equation $\gamma = 1.36(X - 0.26)$. As far as molten slags are concerned, optical basicity values have been shown[30] to correlate well with the sulfide capacities in a number of systems free of reducible transition metal oxides such as FeO and MnO, the expression being $\log C_s = 12.0\ \Lambda - 11.9$ at 1500°C.

Caune *et al.*[31] have suggested that basicity of slags can be measured in a diffusion cell of the type

$$\text{Pt,O}_2 \mid \text{Standard Slag} \parallel \text{Measuring Slag} \mid \text{O}_2\text{,Pt},$$

the EMF being considered to be a measure of $\log a_{O_2{-}}$ in the slag. Here again, the data were correlated with $\log C_s$, the relation being EMF = $(-82.8 \log C_s - 318.3)$ mV.

5.5. Electromotive Force Techniques

Since many molten silicates are ionic conductors, they can be used as electrolytes, but most work has been done using a solid state electrolyte, such as CaO-stabilized ZrO_2, which is ideally suited for the determination of thermodynamic properties in the range of oxygen potentials of greatest interest. Where the electrolyte and liquid slag interact significantly, they have to be kept out of contact with one another, and this can pose one of the main problems in devising the experimental setup.

One of the attractions of this method is that it is so easy to carry out measurements over a range of temperatures, so that partial molar heats and entropies of solution of one component can be obtained from the temperature coefficient of the cell e.m.f. The Gibbs–Duhem equation can then be employed to calculate the activities and partial molar properties of the other component in a binary system, so that the integral molar properties can also be determined, yielding a complete thermodynamic description of the system. A further point is that if readings are taken during both heating and cooling, equilibrium at each temperature can be approached from both sides, confirming that true equilibrium has been attained.

However, great care has to be exercised to ensure accurate results. For example, even with the furnace wound noninductively, the working tube may have to be screened by a molybdenum sheet connected to earth and protected from oxidation by flowing hydrogen. In addition, contact terminals and thermocouple leads may have to be screened and be nonmicrophonic, while a constant mechanical pressure may have to be maintained on the electrodes to insure adequate contact.

The problems and limitations of solid electrolyte galvanic cells have been well dealt with by Rapp and Shores.[32] In addition to the fundamental requirements relating to either electrodes or electrolyte, the reversibility of the cell used in any experiments should be checked by establishing that the e.m.f. is unaffected by temperature cycling, a change of gas flowrate, or the temporary imposition of an external e.m.f.

In cells such as Fe,Fe_xO | electrolyte | slag, Fe, where equilibrium at each electrode involves equilibrium with iron, oxidation of the iron must be avoided by the passage of very pure argon or erroneous results will be obtained. This is much more critical when experimenting with slags than it is with metals and alloys, where a small variation in oxygen pressure only results in a small change in oxide:alloy ratio, without affecting the e.m.f. value. For the same reason, all ceramic components have to be thoroughly degassed and rubber tubing should be avoided, since it is sufficiently permeable to allow contamination of the purified argon.

As an example of the case where the slag itself is used as the electrolyte, the work of Sridhar and Jeffes may be cited. They used simple cells of the type

$$Fe \mid Pb \mid PbO \text{ or } PbO\text{-}SiO_2 \mid Pt,O_2 \mid Pt$$

to measure PbO activities at temperatures from 850 to 1050°C. In practice[33] the cell consisted of an alumina crucible in which liquid lead was in contact with a pure iron electrode, protected by an alumina sheath, and an oxygen electrode, formed by oxygen gas bubbling around a platinum wire inside a small bore alumina tube which just dipped into the electrolyte. This was found to function satisfactorily as long as the end of the platinum wire was held inside the alumina tube, so that it was always in contact with oxygen gas at one atmosphere pressure. This arrangement was also used to study $PbO–B_2O_3$ melts at 900 to 1050°C and the $PbO–P_2O_5$ system, mostly in the range 975 to 1125°C. In each case, the activity of the other component was derived by a Gibbs–Duhem integration, and the partial molar enthalpies and entropies of solution of PbO in the melts calculated from the temperature coefficients of the cell e.m.f. Such a cell can, of course, only be used where the melt is purely an ionic conductor and a suitable material exists for containing both the metal and slag phases.

Subsequent work in the lead silicates and borates has preferred CaO-stabilized ZrO_2 as the electrolyte since it can also function as the crucible, having been shown to be resistant to these melts and not to cause serious contamination of them at temperatures around 1000°C. Kapoor and Frohberg,[34] in particular, have criticized the work of Sridhar and Jeffes on the grounds that atomic oxygen can be transferred through the slag to the electrode in the liquid lead, and, more importantly, that a mixed e.m.f. can arise from reaction between the slag and oxygen around the oxygen electrode leading to the formation of higher lead oxides and, hence, erroneously low e.m.f. values.

Filipovska and Bell[35] have devised an elegant method of determining activities of both FeO and ZnO at 1250°C. A closed system allowed higher p_{Zn} values and hence higher ZnO contents in the slag to be investigated than in previous work. The p_{Zn} was controlled by equilibration with liquid silver, using prior isopiestic measurements of the activity of zinc in silver–zinc alloys at this temperature. A solid electrolyte, CaO-stabilized ZrO_2, which formed the crucible, was used to measure the oxygen potential of the system and the problem of keeping the slag out of contact with the electrolyte was conveniently solved in this instance by the use of an iron container for the slag, which was desired to be iron-saturated. With a Ni/NiO reference electrode, the e.m.f. of the cell reaction

$$(FeO) + Ni_{(s)} = NiO_{(s)} + Fe_{(s)}$$

provides a measure of the FeO activity, once a correction has been made for the thermal e.m.f. of the platinum–iron couple, caused by the use of dissimilar leads. ZnO activities could then be obtained from the known free energy of the reaction

$$Zn_{(g)} + (FeO) = (ZnO) + Fe_{(s)}$$

5.6. *Estimation Techniques*

Slough[36] has reported that a linear relationship exists between the enthalpy of reaction of the crystalline oxides and the ratio of ionic charge to ionic radius for the cation, and this relationship has been used to estimate enthalpies of formation of both silicates and borates.

As far as partial molar free energies are concerned, two models have achieved some measure of success. Richardson's Theory of Ideal Mixing of Silicates is based on the assumptions that the heats of mixing of liquid silicates and, if the anionic constitution of the two binary systems is substantially the same, the configurational entropy of mixing anions, are both negligible, so that the free energy of mixing arises solely from the configurational entropy of mixing cations. Thus "ideal mixing" is not the same as ideal behavior, even though cation mixing is assumed to be random. However, fairly simple expressions have been developed by which the activities of the separate components in various ternary systems have been derived. These compare favorably with experimental values for limited ranges of composition around the metasilicate in cases where the cations have similar depolymerizing tendencies, as elucidated by the polymerization models of Masson.

The other model is based on assuming that all oxides, including SiO_2 and other network-formers, ionize simply to form cations and O^{2-} anions, so that on mixing two binary melts to form a ternary, no anionic mixing is involved. Cationic mixing is again assumed to be random and so the expressions developed are exactly similar to those derived from regular solution theory, but conformity to them should not be taken to indicate regular solution behavior. In spite of its neglect of structural realities this model has been shown to have a surprisingly wide applicability, not just to ternaries but also to quaternaries, and to melts involving alumina[37] and titania.[38]

Such models can be used to complement and supplement experimental studies in various ways. The interpolation and extrapolation of data can be made easier and more accurate, thereby reducing the number of experimental determinations necessary to give adequate coverage of the system. Also, the preliminary calculation of approximate activity values means that starting compositions fairly close to the final equilibrium compositions can be employed and the time necessary for equilibration reduced accordingly.

References

1. G. Sommer and P.R. Jockens, *Minerals Science & Engineering* 1971, 3, 3; J.H. Welch, *J. Sci. Instr.* **38,** 402 (1961).
2. M.A. Rouf and H.B. Bell, *Pakistan J. Sci. & Ind.* **13,** 179–184 (1970).
3. T.R. Meadowcroft and F.D. Richardson, *Trans. Farad. Soc.* **61,** 54–70 (1965).
4. C.R. Masson, W.D. Jamieson and F.G. Mason, *Physical Chemistry of Process Metallurgy,* Inst. Mining & Metallurgy, pp. 223–231 (1973).

4a. E. Franks and D. Inman, *J. Appl. Electrochem.* **1,** 73 (1971).

5. Y. Waseda, *The Structure of Non-Crystalline Materials, Liquids and Amorphous Solids,* McGraw-Hill, New York, 1980.
6. D.R. Gaskell, A. McLean, and R.G. Ward, *Trans. Farad. Soc.* **65,** 1498–1508 (1969).
7. H. Towers and J. Kay, *Trans. Brit. Ceram. Soc.* **49,** 1 (1956).
8. J.O'M. Bockris and D.C. Lowe, *J. Sci. Inst.* **30,** 403–405 (1953).
9. E.F. Riebling, *Rev. Sci. Inst.* **34,** 568–571 (1963).
10. D.W.F. White, *Trans. ASM* **55,** 757 (1962).
11. T.B. King, *Trans. Soc. Glass Tech.* **35,** 241–259 (1951).
12. H. Towers, M. Paris and J. Chipman, *Trans. A.I.M.E.* **197,** 1455–1458 (1953).
13. P. Perron and H.B. Bell, *Trans. Brit. Ceram. Soc.* **66,** 347–365 (1967).
14. M. Sasabe and Y. Kinoshita, *Trans. Iron Steel Inst. Jpn.* **20,** 801–810 (1980).
15. P.J. Spencer, National Physical Laboratory Report Chem. 21, February 1973.
16. I.D. Sommerville and H.B. Bell, "Molten Slag Systems—Phase Equilibria and Chemical Activities of Components," "*Schlacken in des Metallurgie,*" Verlag Stahleisein mbH., Düsseldorf, 1983. Ed. D. Janke.
17. O. Kubaschewski, *High Temp.-High Pressures* **4,** 1–12 (1972).
18. W. Slough and G.P. Jones, National Physical Laboratory Report Chem. 31, August 1974.
19. A.W. Richards and D.F.J. Thorne, "Physical Chemistry of Process Metallurgy" Part 1, p. 277–291. Edited by A.R. St. Pierre, Interscience Publishers, New York (1961).
20. G.R. Belton, U.V. Choudary and D.R. Gaskell, "Physical Chemistry of Process Metallurgy: the Richardson Conference", Inst. of Mining and Metallurgy, London (1974), pp. 247–253.
21. G.R. Belton, H. Suito and D.R. Gaskell, *Met. Trans.* **4,** 2541–2547 (1973).
22. A. McLean, *High Temp.-High Pressures,* **6,** 21–31 (1974).
23. A.S. Venkatadri and H.B. Bell, *Trans. Met. Soc. AIME,* **245,** 2319–2323 (1969).
24. J. Cameron, T.B. Gibbons and J. Taylor, *Iron Steel Inst., London* **204,** 1223–1228 (1966).
25. R.A. Sharma and F.D. Richardson, *Iron Steel Inst., London* **200,** 373–379 (1962).
26. D.J. Zuliani, M. Iwase, A. McLean and T.R. Meadowcroft, *Can. Met. Quart.* **20,** 181–187 (1981).

27. R. Stermsek and K.W. Lange, *Can. Met. Quart.* **20,** 189–197 (1981).
28. C. Wagner, *Met. Trans.,* **6B,** 405–410 (1975).
29. J.A. Duffy and M.D. Ingram, *J. Non-Crystalline Solids* **21,** 373–387 (1976).
30. J.A. Duffy, M.D. Ingram, and I.D. Sommerville, *J. Chem. Soc., Faraday Transactions I,* **74,** 1410–1419 (1978).
31. E. Caune, M.G. Frohberg and M.L. Kapoor, *Archiv. Eisenhüttenwesen,* **49,** 271–276, 333–338 (1978).
32. R.A. Rapp and D.A. Shores, "Physicochemical Measurements in Metals Research," Volume 4. Part 2, 123–192. Edited bv R.A. Rapp, Interscience Publishers, New York (1970).
33. R. Sridhar and J.H.E. Jeffes, *Trans. Inst. Mining & Metallurgy* **76,** C44–C50 (1967).
34. M.L. Kapoor and M.G. Frohberg, *Can. Met. Quart.,* **12,** 137–146 (1973).
35. N.J. Filipovska and H.B. Bell, *Trans. Inst. Mining & Metallurgy* **87,** C94–C98 (1978).
36. W. Slough, National Physical Laboratory Report Chem. 25, October 1973.
37. I.D. Sommerville, I. Ivanchev and H.B. Bell, "Chemical Metallurgy of Iron and Steel," International Symposium, Sheffield, 1971, pp. 23–25. Iron and Steel Institute (1973).
38. I.D. Sommerville and H.B. Bell, *Can. Met. Quart.* **21,** 145–155 (1981).

4

Molten Salt Spectroscopy

Trevor R. Griffiths

1. Introduction

Spectroscopy has long been recognized as a most valuable technique which yields considerable chemical information concerning both solutes and solvents. It is therefore not surprising that molten salts have been examined by every branch of spectroscopy. Not every spectroscopic technique is readily applied to molten salts and thus some approaches have received less attention than others. This is largely due to practical difficulties. For example, spectrophotometers suited to high-temperature measurements are but rarely offered by manufacturers and, in consequence, researchers who have persisted in this field often have had to devise and manufacture specialized attachments or modifications for their instruments. Sometimes the basic apparatus is not suited to high-temperature studies: the space between the pole pieces of a standard nuclear magnetic resonance spectrometer is certainly not large enough generally for a high-temperature furnace to be located around the sample. However, W.W. Warren Jr. and his co-workers at the Bell Telephone Laboratories are recording spectra successfully up to 1000°C, albeit with a specialized spectrometer. Electron spin resonance fused salt data are almost nonexistent, though Swanson[1] has reported esr studies on $MnCl_2$ in LiCl–KCl eutectic, and Kukk[2] has described some preliminary work on transition metal halides in fused halide solvents. The esr technique will not be discussed further here.

At the present time technical advances, to a certain extent in miniaturization, which enables small efficient furnaces to be constructed, but more particularly in data capture, have the potential to give molten salt spectroscopy, in all its branches, a new lease on life. For example, data capture now facilitates the computer manipulation of spectra and this offers the opportunity to derive much

Trevor R. Griffiths • Department of Inorganic and Structural Chemistry, The University, Leeds LS2 9JT, England

more information from molten salt spectra than had previously been possible. Details have recently been described.[3] In addition, a new spectroscopic technique also has been presented[4] which yields much new and useful information on the nature and rate of the corrosion of metals by molten salts experiencing a heat flux. Brief descriptions of both of these developments are included in this Chapter.

The spectroscopic techniques which have received the most attention over the years are those which cover the electromagnetic spectrum from the ultraviolet to the infrared, including Raman studies. Even these applications are relatively few and, hopefully, this account will encourage others to employ such techniques and methods. A continual source of interest to this author is that while conventional spectroscopic procedures and practices can be employed, there is always the challenge to devise ways of preparing, and maintaining, uncontaminated samples. These are generally small volumes of molten salts containing some solute at an accurately known concentration.

Each section of this chapter is organized in the following manner when appropriate. Firstly, there is a discussion in general terms. Secondly, any particulars pertinent to the various techniques are given, starting with those used for making measurements in the ultraviolet-visible region and then those concerned with the infrared and Raman region. Finally, any points relating to nmr measurements are given.

2. *Scope of Fused Salt Spectroscopy*

A valuable property of those molten salts generally considered as solvents, *viz.*, alkali and alkaline earth halides, nitrates, and sulfates, is that they are optically transparent over a wide wavelength range. Indeed, the wavelength limit of available spectrophotometers in the far ultraviolet region is usually the limiting factor for fluoride, chloride, and sulfate melts. In the infrared region molten alkali halides, nitrates, and sulfates are often more transparent than water, and thus many transitions masked by water or some other solvent absorption bands can now be measured.

The uv limit at which a molten salt may no longer be suitable as a solvent is best gauged from its uv cut-off, the wavelength at which the absorbance is unity for the salt in a cell having the path length of interest. Of course, if the rate of absorption increase of this fundamental edge is found to be shallow, then data may be recorded with reasonable confidence below the uv cut-off value. The uv cut-off will, however, depend upon the temperature of the melt. As a general rule, this value will shift toward the infrared region by 1000–2000 cm^{-1} for each 100°C rise in temperature. The limit in the infrared region needs to be determined by experiment. This is generally done by recording a base-line solvent reference spectrum of the pure melt with air in the reference beam of the spectrophotometer.

In more recent years, the development of lasers has led naturally to the revitalization of Raman spectroscopy. The spectrometers and illuminating systems and techniques which preceded this advent will not be discussed here, but it should be pointed out that besides simplifying the measurement of the Raman spectra of fused salts, many new systems could now be studied, including colored solutions.

3. *Spectrophotometers*

3.1. *UV-Visible-Near Infrared Spectrophotometers*

As indicated above, commercial spectrophotometers are not designed for high-temperature samples, with one exception which will be discussed later. The use of commercial instruments causes problems because of the radiation emitted by the sample and its container, and the furnace. Two problems commonly arise. Firstly heat, largely from the furnace, can affect the dimensions and electronics of the instrument, particularly if the cell compartment is small and the amount of thermal lagging around the furnace is limited. The monochromator may warm up with time, and this affects the accuracy of the instrument. Some instruments whose electronic components may bring about this are supplied with thermostatted monochromators. Alternatively or in addition to thermostatting, the passage of cooling water through copper pipes attached to the exterior of the furnace or to the interior of the cell compartment will essentially overcome this problem. The latter cooling method may be necessary if the configuration of the spectrophotometer is such that the sample is located next to the detector. The characteristics of the detector, particularly its noise, are adversely affected by slight temperature increases. The temperature range permissible for a detector is generally found in the manufacturer's literature.

The second and major problem is that of the radiation emitted in and around the region of the light beam of the spectrophotometer reaching the detector. Early work[5,6] showed emission of radiation (from samples up to 250°C) did not constitute a problem for single beam, manual spectrophotometers covering the visible range. Molten salts at much higher temperatures behave as almost perfect black-body emitters, closely fitting the Planck equation for the black-body spectrum. The thermal emission spectra of a black-body at different temperatures occur as a set of distribution curves, each of which shifts towards the region of shorter wavelengths as the temperature rises, according to the well-known Wein's displacement law. The intensity of the radiation between any given wavelength limits increases steadily as the temperature rises. Thus phototube and photomultiplier detectors, which are sensitive to radiation in the visible region, can be swamped by white-hot samples. However, the effects may be noticed at lower temperatures, as shown by a change in the signal (monochromatic beam) to noise (emitted radiation) ratio, which diminishes quickly in conventional spectrophotometers in the range 250–500°C.[7] A change in the sequence of the major components in a spectrophotometer can ameliorate the situation.

The simplest sequence, which applied to the instruments first used in molten salt spectroscopy, which are manually operated, is:

Light source → Monochromator → Sample → Detector

(Scheme A)

Here the detector receives the complete spectrum of emitted radiation as well as the transmitted monochromatic beam. The maximum sample temperature that can be employed is about 500°C.

Automatic recording spectrophotometers will permit the use of higher sample temperatures. In this case the sequence is:

Light source → Monochromator → Chopper → Sample → Detector
→ A.C. Amplifier → Recorder

(Scheme B)

In this scheme the light beam is pulsed before passing through the hot sample. The detector receives all radiation emitted by the sample, its container, and the furnace, that is, the complete spectrum of thermal radiation, plus pulsed, monochromatic light, at the wavelength of interest. The detector signal thus consists of both an ac and dc output, and only the ac current is amplified and sent to the recorder. In principle, therefore, this configuration should work at all temperatures, but in practice the maximum sample temperature is restricted to around 750°C. This is because the detector, being adjacent to the sample, is heated and its noise level is increased. Some reduction is possible by introducing water-cooling and heat-absorbing glass filters between the furnace and the detector (but with the consequent absorbance of ultraviolet radiation limiting the recording to visible spectra only). However, the main restriction is due to the relative magnitudes of the ac and dc signals. When the latter is large, the net effect on the ac signal is to increase markedly its noise level, and hence resolution is lost.

There are two schemes which permit operation at higher temperatures, namely,

Light source → Sample → Monochromator → Detector

(Scheme C)

Light source → Sample → Chopper → Monochromator → Detector →

A.C. Amplifier → Recorder

(Scheme D)

Acceptable absorption spectra have been recorded up to 1050°C, using Schemes C[8] and D.[9] By interposing the monochromator between the sample and the detector, the emitted radiation reaching the detector is limited to that of the wavelength setting of the monochromator. Thus, when compared with Schemes A and B, at their maximum permissible temperatures, the detector here encounters a significant reduction in the emitted radiation.

Scheme D, however, does produce one difficulty—the thermal radiation is pulsed at the same frequency as the transmitted beam, and it is not now discriminated against by the ac amplifier and is sent to the recorder as part of the signal. This can be allowed for since, as noted above, molten salts are good black-body emitters. Using Planck's equation for black-body radiation reasonably accurate estimates have been calculated[9,10] of the ratio I_E/I_O, where I_E is the intensity of the sample and furnace radiation, and I_O is the intensity of radiation of the light source (a tungsten lamp). An example[11] reproduced here illustrates the relative magnitudes of this effect.

Assuming a sample temperature of 2400°C, I_E/I_O has the values 0.02 and 0.09 at 1500 nm and 2500 nm, respectively. Suppose, however, that a transition metal ion, dissolved in a molten salt at 1000°C has, at 1500 nm, an absorbance of unity. It thus absorbs 90% of the incident light at this wavelength. The intensity of the thermally emitted radiation under these circumstances constitutes 20% of the signal. Since for a given sample temperature the ratio I_E/I_O is, of course, wavelength dependent, and since the spectrophotometer does not distinguish the (monochromatic) thermal radiation from the transmitted radiation, a correction has to be applied to the recorded spectrum at every wavelength, a tedious business.

We now discuss the one commercial uv–visible–near ir spectrophotometer that achieves essentially complete removal of the emitted radiation, the recorder response comprising only the transmitted radiation. The scheme is:

Light source → Chopper → Sample → Monochromator → Detector →
A.C. Amplifier → Recorder

(Scheme E)

This configuration is employed in the Applied Physics Cary 14H spectrophotometer, a tube operated instrument which has now been superseded by the Varian Cary 17H and 17DH models (the latter is fitted with digital output). The optics have not been altered, the 17 series now uses solid state circuitry and is capable of being interfaced with computing equipment. The optical arrangement is shown in Figure 1. Scheme E may be recognized in atomic absorption spectrometers and is often used in infrared instruments. Emitted radiation is eliminated because now the detector receives pulsed (ac) monochromatic transmitted radiation and continuous (dc) monochromatic emitted radiation. The latter, for hot samples, is of such significantly lower intensity than the former, that the reduction in signal-to-noise ratio compared with room temperature spectra, is scarcely detectable. An interesting early test[12] was the location of a tungsten filament bulb in the sample light beam. The "spectrum" was recorded before and during the passage of current, when the hot filament behaved as a sample (and furnace) at around 2200°C. A slight increase in noise level (which could be essentially removed with a small increase in the pen damping setting[13]) thus demonstrated that the instrument discriminated essentially 100% against emitted radiation. Thus the spectra of samples well above 2200°C are capable of accurate measurement with this instrumental sequence.

Scheme E together with Schemes C and D, also permit spectra to be recorded without the tiresome necessity of making sure that a sample housing modified to accept a furnace is light-tight. It has been suggested that nearby fluorescent lights could be in phase with the chopper and thus cause erroneous spectra. In this unlikely event such radiation could only enter the instrument as stray light, and the optical arrangement employed very effectively removes stray light—this author has never experienced this effect.

Scheme E does, however, have two slight disadvantages over the more conventional Scheme B. The first is that by chopping the light before it enters the monochromator the light intensity at the detector is reduced by a factor of approximately two. This is readily overcome by replacing the tungsten lamp with a quartz halogen lamp. At the normal operating voltage the latter is approximately equivalent in output to a tungsten lamp. The light intensity loss can be more than compensated by over-running the lamp slightly, a practice which only marginally reduces its operating life. (The facility to vary the applied voltage is useful when small samples or specialized sample holders which reduce the normal cross-section of the light beam are employed, thus enabling normal slit widths to be used.) The second disadvantage is that the sample is always exposed to the total emission of the light source and some samples may be photosensitive. The author has not experienced this with his molten salt spectra but has noted it with studies of the charge-transfer-to-solvent spectrum of the iodide ion in nonaqueous media, and in the spectra in single crystals of lattice defects generated by particle bombardment: here optical bleaching occurred.

A Unicam SP700 spectrophotometer has had its optics reversed so that it can be operated in the configuration of Scheme E.[14] It is understood that the optics were properly adjusted so that the reference and sample beams came to a focus where the cells containing the molten salts were located. A detailed description of this conversion has yet to be published.

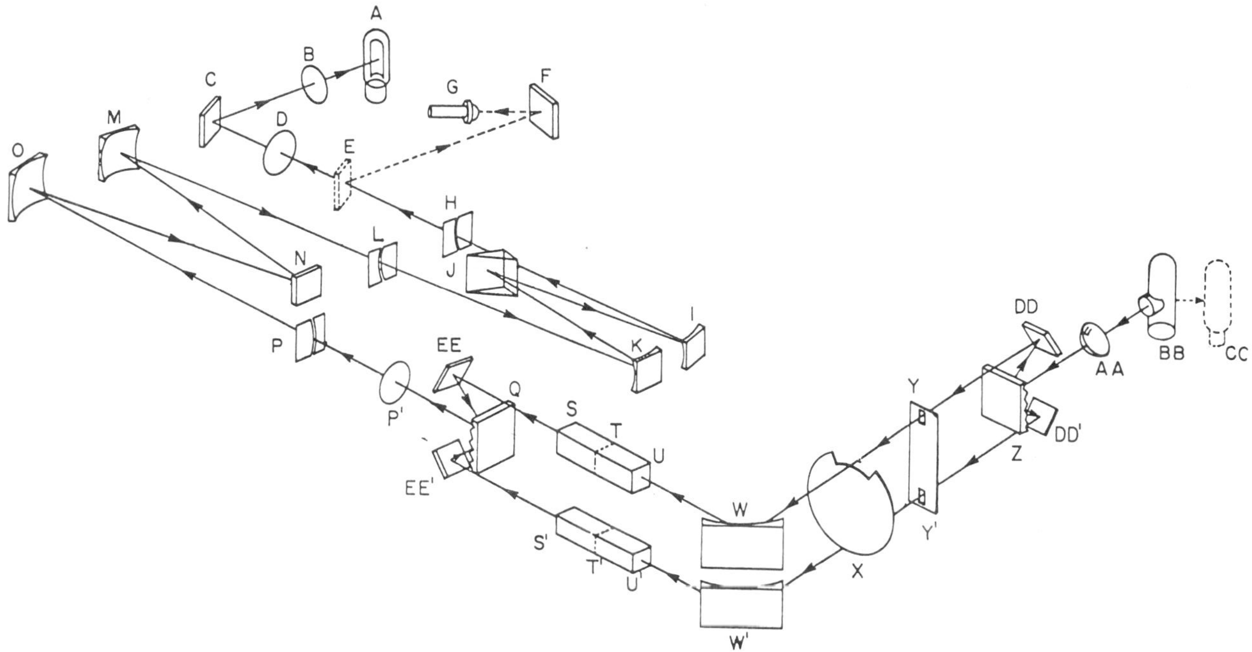

FIGURE 1. Optical arrangement of the Cary 14H (and 17H) spectrophotometer. BB,CC, light sources; AA, condensing lens; X, beam splitter; DD,DD′, focussing mirrors; YY′, screens; X, chopper; WW′, focussing elements; EE,EE′, mirrors; U,U′, image of vertical slit; T,T′, horizontal slit image; S,S′, image of vertical aperture; Q, beam combiner; P′, cylindrical lens; P,L,H, monochromator slit system; O,M, grating collimator mirrors; N, grating; K,I, prism collimator mirrors; J, prism; E, movable mirror; D,B, focussing lenses; C, flat mirror; F, focussing element; G, PbS detector; A, phototube detector. Courtesy of Varian Associates.

3.2. *Infrared Spectrophotometers*

3.2.1. *Transmission Spectroscopy*

Studies of the infrared spectra of molten salts, also technically challenging, do not experience the same instrumental difficulties that accompany work with most commercial uv–visible–near infrared spectrophotometers. Most commercial near-infrared (and Raman spectrometers, as we shall see later) require little adaptation for recording the spectra of molten salt samples, provided they have a sufficiently large and accessible sample compartment. The early workers in this area overcame the major difficulties, but it is surprising that the number of workers has declined. For technical reasons the measurement of spectra as high as 1000°C has not generally been possible. In 1969 Devlin[15] remarked that the main contributions to the field seemed to divide, both as to content and origin, into two groups: his own group and that of Hester and co-workers. Since then even these groups have changed direction. There is still much work to be done as the topic is by no means exhausted.

To avoid the direct radiation from a hot sample contributing to transmitted infrared radiation, many commercial near-infrared double-beam spectrophotometers employ a rotating chopper to recombine sample and reference beams as they exit from the sampling compartment (Scheme D). Introduction of a new chopper between the infrared radiation source and the sampling compartment essentially changes the configuration to Scheme E. Some manufacturers sell such a prechopper as an accessory, but since the chopping frequency needed for use with slow-response infrared detectors is low (*ca.* 10–15 Hz), simple homemade beam-splitters will suffice. A reciprocating flag type comprising a slotted, front-surface-aluminized metal or glass plate has been recommended.[16,17]

To investigate the far infrared region (*ca.* 30–400 cm^{-1}), the above modifications can be implemented to commercial spectrophotometers with minor changes as necessary. However, the introduction of a prechopper is not required when interferometers of the Michaelson type are used, as they are already incorporated. To date very little molten salt far-infrared studies have been reported, largely because the usual polyethylene and polypropylene lens and window materials cannot withstand the heat levels involved. Work employing diamond windows has been announced,[18] but the limiting useful temperature for diamond is around 600°C.

Recognition that the radiation emitted by molten salts approximates closely to blackbody radiation at the same temperature was the basis for another method for studying high-temperature infrared spectra. Wilmshurst[19] used the optical system shown in Figure 2, where the light emanating from the sample is compared with that from a blackbody at the same temperature. When these emissivity studies were compared with reflectivity data, similar results were found for the derived absorption spectra from both techniques.[19] Later Kozlowski[20] also used

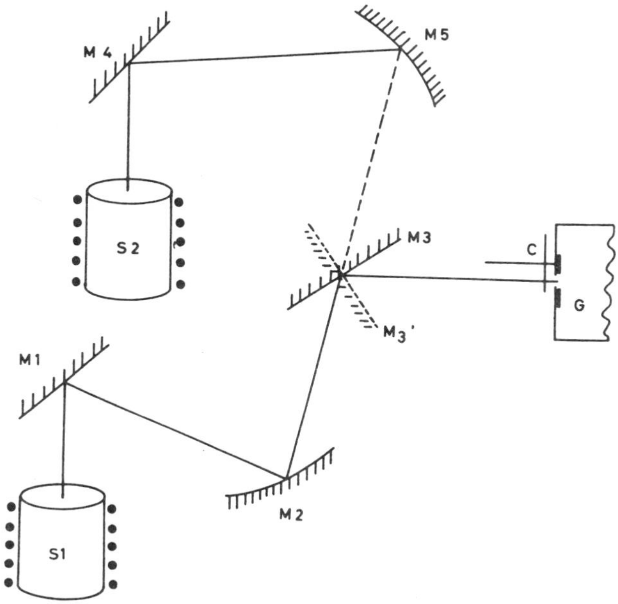

FIGURE 2. Optical arrangement used for infrared emission spectroscopy of molten salts. S1, sample; S2, black body emitter; M, mirror; C, chopper; and G, monochromator. Courtesy of Plenum Press.

emissivity studies for molten salt spectra, and though the technique is not easy experimentally, it merits further attention and development.

Fortunately, modern ratio-recording infrared spectrometers do now permit the measurement of absolute band intensities and allow tests of Beer's law in the infrared region, thereby raising the performance of these instruments to the level of visible spectrophotometers. It is therefore with such instruments that new entrants to this field would be best advised to commence.

3.2.2. Reflectance Spectroscopy

Reflectance spectroscopy is another technique used by the early workers to overcome the containment problems of holding an often corrosive molten salt in an infrared beam. It was shown that spectra can be obtained[19,21,22] and that they compare well with comparable direct absorption measurements.[23] The basis of the experimental set-up of Wilmshurst and Senderoff[21] is shown in Figure 3. Absorption spectra were derived from their reflectance data, using the method of Robinson and Price.[24] Though container problems are lessened because crucibles of any suitable material may be used, considerable spectrometer changes

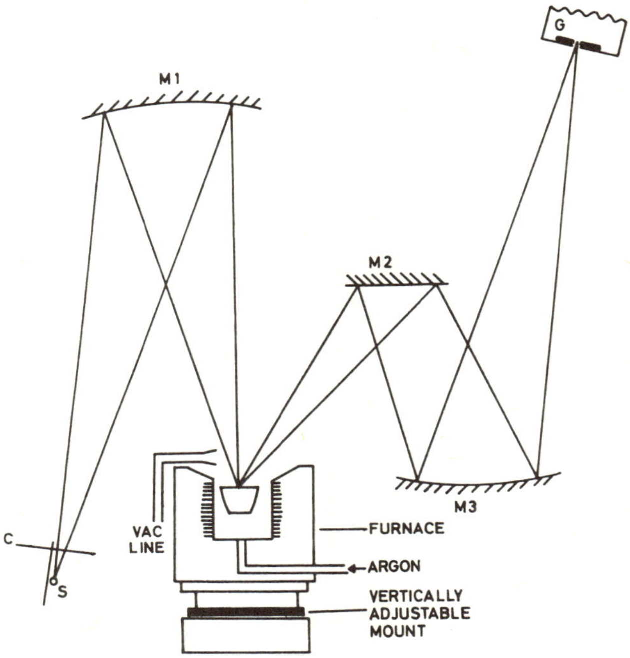

FIGURE 3. Schematic diagram of an optical arrangement used to obtain infrared reflectance spectra of molten salts. S, source; C, chopper; M, Mirror; and G, monochromator. Courtesy of Plenum Press.

are necessary, and thus this technique is little used now. Other difficulties were the elaborate methods necessary for interpretation and the poor precision available.

3.2.3. Attenuated Total Reflectance (ATR) Spectroscopy

Before the advent of ratio-recording infrared spectrophotometers it was recommended[15] that ATR measurements were needed to complement transmission studies because ATR makes possible (a) reproducible and adjustable pathlengths, (b) short effective pathlengths in the 1 μ range, and (c) the virtual elimination of band-distorting reflection problems. For completeness therefore, a brief account of the technique is included.

The principle is that the infrared beam passes from an optically dense medium (the ATR prism, Figure 4) to the less dense (lower refractive index) sample.[25] Total reflection takes place at the interface for frequencies where the sample is nonabsorbing, on condition that the angle of incidence θ and the refractive index n at the interface are such that $n < \sin\theta$. The energy of the reflected beam is lowered by absorption in the surface layers of the second medium (the molten salt sample) when this has a nonzero molar absorbance ε_M at the wavelength used. The reflectance spectrum exhibits attenuation at those frequences where $\varepsilon_M \neq 0$, and the profile resembles closely a transmission spectrum, provided parameters such as incident angle and prism refractive index have been maximized for that purpose.[25,26] Temperatures below 400°C have been employed using a silicon prism.[27] Irtan materials have some advantages[22] but a protective coating is needed to prevent attack by molten salts. Precise absolute intensity and band profile studies require specialized optics to obtain the necessary reflection measurements over a range of carefully defined angles of incidence, or so it was claimed in 1966.[28] A recent review by Wilks[29] on internal reflection spectroscopy points out that if the top of the prism is cut off to form a trapezoid with all faces polished, and a beam is sent through the remaining portion of the prism face at an angle greater than the critical angle, the beam is reflected from one parallel face to the other until it reaches the angled face perpendicular to it. Striking this surface at less than the critical angle, it exits. Today, using multiple reflector plates, the angle of incidence can be set well away from the critical angle, thereby eliminating spectral distortion. And, though there are some differences between internal reflection and transmission spectra, resolution, band shapes, and band positions are identical. Current technical developments have made ATR a very promising and viable technique for obtaining, essentially, infrared transmission spectra through very thin films of melts; but at the present time it is still largely neglected by molten salt chemists.

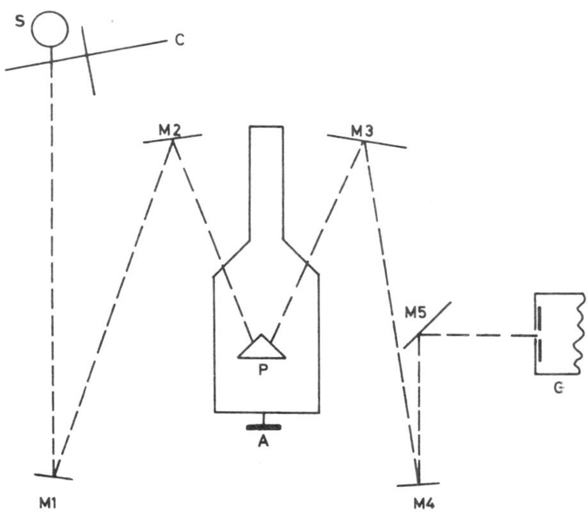

FIGURE 4. Optical path for ATR molten salt spectroscopy. S, source; C, chopper; M, mirror; P, ATR silicon prism; A, sample container; and G, monochromator. Courtesy of Plenum Press.

3.3. Raman Spectrophotometers

Raman and infrared spectra are complementary and, usually, both are required to obtain a complete set of vibrational frequencies in order to assign a molecule (or ion) to a certain point group and to determine force constants. Until the introduction of the laser to Raman spectroscopy in 1962,[30] and the use of continuous gas lasers[31,32] with photoelectric recording of spectra,[32,33] the mercury arc, particularly the Toronto arc,[34] was the standard source of radiation. Hester has given a brief account of the arc and discharge lamps previously used.

The laser, with its narrow (usually 1–2 mm diameter) well-collimated beam of radiation, enables many of the previous experimental difficulties to be overcome. A small hole is required for the beam to enter the sample furnace, plus another for the scattered light to exit. The beam can be focused to a diffraction-limited spot of dimensions comparable to its wavelength, and unless stimulated Raman effects are required, the continuous gas lasers are currently more useful than pulsed lasers.

Commercial Raman spectrometers fitted with lasers are generally ideal for high-temperature studies. A prechopper is not necessary before around 1200°C is reached by the sample if an argon laser exiting at the blue end of the spectrum is employed, because there is relatively very little direct emission in the short wavelength end of the visible spectrum from such samples. (Photoelectric detection is assumed—photographic plates would be blackened by the necessary long exposure to the thermal emission.) With He–Ne lasers exciting in the red end at 632.8 nm, the prechopper is needed for hot samples, together with signal detection by red-sensitive photomultipliers. In contrast to most infrared detectors, the fast response characteristics of photomultipliers make it desirable[35] to pulse in the 500–1000 Hz range.

Several commercial laser-Raman spectrometers are available, with argon and/or helium-neon lasers for continuous recording of spectra. Instruments combining two spectrometers in series (double monochromator) are recommended[36] to minimize stray light. A schematic diagram is shown in Figure 5 which represents the system used by Maroni and co-workers.[36] A Spex Industries 1400 double spectrometer was linked to a Spectra Physics 125 continuous He–Ne gas laser (nominal power 50 mW at 632.8 nm). The detector was the cathode of a Hamamatsu R136 photomultiplier tube cooled to -50°C to decrease tube shot noise. The prechopper operated at 400 Hz and the photomultiplier signal was fed to an A.C. lock-in amplifier linked to a phase-angle voltmeter. The reference signal for the amplifier was provided by a photo-resistive detector, which received a small fraction of the laser radiation reflected from the front surface of the prism. Spectral slit widths were between 5 and 10 cm^{-1}. Spectra were normally recorded at a scan rate of 2 cm^{-1} min^{-1} with a 50 s time constant, or 5 cm^{-1} min^{-1} with a 10 s time constant, and plotted on a strip-chart recorder.

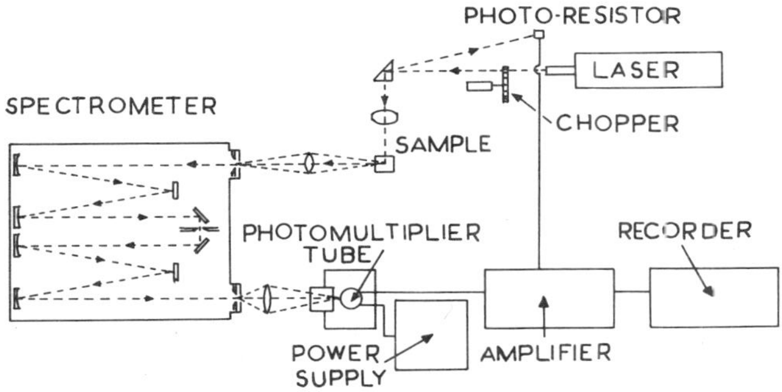

FIGURE 5. Schematic diagram of a laser-excited Raman spectrophotometer.

3.4. Nuclear Magnetic Resonance Spectrometers

Some modifications are necessary to commercially available nmr instruments to permit the recording of the chemical shifts of molten salts at high temperatures. Early work had shown that it is difficult to adapt the Bloch crossed coil probe to temperatures above 500°C, due to the difficulty of maintaining precise geometry. A single coil with an r.f. bridge can be used, though the sensitivity is now slightly reduced. The nmr signal is obtained through the bridge, which is adjusted to a slightly off-balance condition. A major advantage, however, is that more space is available for the heating and cooling systems, and for r.f. shielding. An account of some specialized probe and furnace combinations that have been successfully employed will be given in the section on furnaces.

4. Furnaces for Spectrophotometers

A wide variety of furnace designs for use in molten salt spectrophotometry has been described over the years. Excluding for the moment those furnaces designed for nuclear magnetic resonance and electron spin resonance studies, there are several features common to furnaces for spectrophotometers.

4.1. General Features

In the first place, unlike most furnaces, uniform lagging around the hot zone is not possible: opposing light ports are necessary (except in Raman studies where two at right angles are normally required). Hence considerable care in the design, construction, and operation of such furnaces is necessary to provide and maintain isothermal conditions in and around the region of the sample through which the light beam is passed.

Second, although this may be overcome more easily the larger the furnace, spectrophotometers generally are designed not with parallel optics but such that the light beam comes to a focus at a particular point, and this is where the molten salt sample should be located. It is not usually possible, whichever scheme is used, to increase the linear distance between, say, the monochromator and detector in order to insert a well-lagged isothermal optical furnace. And in double-beam instruments the separation between sample and reference light paths is small also, around 10 cm, and this too limits the practical diameter of the furnace. Thus the dimensions of the furnace are limited by the size of the sample compartment and the distance between the sample and reference beams. Since the majority of applications require spectrophotometers for routine, room temperature, analytical and clinical studies there is little incentive to provide instruments that are sufficiently versatile for the molten salt research community.

Third, since the difficulty of maintaining a sample and reference molten salt under identical temperature (and atmospheric) conditions is thereby compounded, many workers have opted to record all their spectra against air at room temperature as reference, subtracting from their recorded spectra the spectrum of an appropriate reference molten salt solution also recorded against air. (Using digitized spectra this chore is minimized.) Spectrophotometers are thus being used essentially as single beam instruments, and the assumption is made that the electronics of the instrument does not drift between recording sample and reference spectra.

Fourth, because the cells containing the molten salts are liable at times to break or crack and leak, the majority of furnace designs for molten salt work envisage that the core can be readily removed for cleaning and spilled salt will not be able to reach any components it might damage.

Finally, though all furnaces contain insulation, the material chosen is not generally important. Fire-brick, bubbled alumina, asbestos, vermiculite, powdered zirconia, diatomaceous earth, and Kaowool have all been used. Because of the space restrictions that commonly apply, a more general rule is that if the outer casing gets more than hand-hot, external water cooling must be installed.

4.2. Furnaces for UV-Visible-Near Infrared Spectrophotometers

4.2.1. Isothermal Furnaces

Perhaps the best designed furnace for use with uv–visible–near infrared spectrophotometers is that of Boston and Smith.[37] It was originally designed for use up to 400°C, because it had an aluminum core, and it is compact enough to fit in the cell compartments of most commercial spectrophotometers. With an alternative core material higher temperatures may be attained. The design is

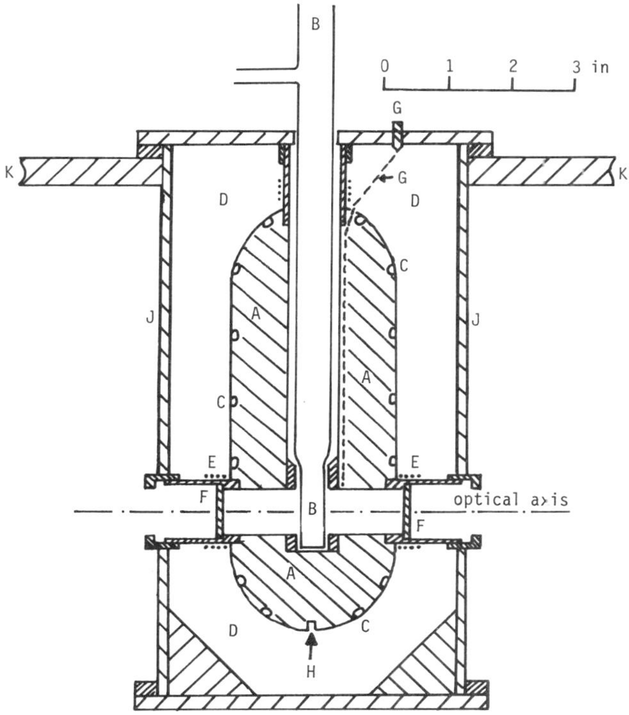

FIGURE 6. Cross-section of the 400°C aluminum block furnace. A, aluminum heat sink; B, optical cell; C, recessed heating elements; D, insulation; E, guard heater; F, silica window; G, thermocouple in sleeve; H, well for controlling thermocouple; I, outer jacket.

illustrated in Figure 6. The main feature of the design is that the heat input into the aluminum heat sink is uniform. The metal-sheathed heating wire is recessed into the surface of the block, the position of the channels having been carefully calculated to provide uniform heating. There are secondary, independent windings around each side arm which, together with optically flat silica windows, prevent convectional chimney effects and other heat losses through the light ports. However, it is a tedious and time-consuming exercise to adjust and control the power supplies to these windings and the main heating wire so that the vertical and horizontal thermal gradients in the area of the light beam are minimized. Boston and Smith[37] reported that the maximum temperature difference in the region of the furnace occupied by the molten salt in the cell was 0.5°C (over about 10 cm vertically), and less than 0.2°C in the window regions, and they further suggested that temperature differences in a melt should be less than this due to thermal convection.

We have constructed a furnace based on the Boston and Smith[37] design, but with certain simplifications.[38] The main metal-sheathed heating wire was connected in series to the secondary windings around the side arms, the number of windings was gauged initially from experience, and adjusted as necessary. This compromise meant that the furnace could be maintained to within 1°C of the set temperature and that differences in temperature between the aluminum block, as determined by the thermocouple in the well alongside the cell, and the molten salt, were less than 1°C for temperatures up to 200°C. The thermal gradient along the vertical axis of the furnace at this temperature in the melt was determined as 0.95°C cm^{-1}. The corresponding horizontal gradient in the molten salt located in the light beam was less than 0.2°C cm^{-1}. These parameters are adequate for almost all molten salt studies. We did not employ the water-cooling fitted to the original design,[37] partly because our Cary 14H spectrophotometer has a large cell compartment, and partly because early tests had shown that the brass outer casing did not get hand-hot (and soldering or brazing cooling pipes on to the casing is not technically easy). However, we would anticipate that cooling will be necessary when this design is extended to higher temperatures on replacing the aluminum block with, say, stainless steel. (Boston and Smith[39] have used a furnace containing a solid silver block: it reached an operating temperature of 600°C in a few minutes, but since the likelihood of anyone in the future being able to (afford to) construct such a furnace is remote, its details are not included here.)

A copper core furnace which permits simultaneous heating of sample and reference compartments has been described[40] for the Unicam SP 700 spectrophotometer, and it can readily be installed and removed from the standard instrument. Temperature control to within ±1°C is reported at 600°C, though if much work is anticipated above 500°C a stainless steel block is recommended.[41] Temperature uniformity between the two cell compartments is said to be within ±0.5°C. However, like most designs, the thermal gradients, particularly the vertical gradient in the region of the light beam, are not given. It is our experience that good temperature control of a furnace does not necessarily mean the absence of thermal gradients. This design uses six cylindrical heaters inserted into holes, arranged symmetrically around the sample and reference cell positions. All round water-cooling effectively eliminates significant heat transfer to the spectrophotometer. Since the Unicam SP 700 utilizes the configuration shown in Scheme B the only effect of thermal radiation is to heat the detector and, at the furnace temperatures involved, the loss in resolution is probably unimportant in view of the broadening of bands at elevated temperatures. This furnace was an improvement on an earlier design[42] for the Unicam SP 700. It would probably be difficult to improve upon this new design for most molten salt studies with this spectrophotometer.

An example of ingenuity when apparatus was limited has been shown by Johnson and co-workers.[43] They had only a Beckman DK2 spectrophotometer available, which has a very small sample compartment (12 cm long, 7 cm deep,

9 cm wide), making the replacement of the normal cell-holder by a furnace almost impossible. They therefore used mirrors to reflect the light beam out of the compartment, through the furnace and back to the detector (Figure 7). The furnace had a stainless steel core, and was similar in many details to that of Bailey and McIntyre[40] described above. Temperature stability within the melt at 450°C was ± 1°. To record spectra between 500 and 800°C, 500°C being the limit for spectrophotometers with the component arrangement of Scheme A, the detector was protected from thermal radiation with heat filters, which still permitted measurement of visible spectra.

However, they did not discuss whether the light rays in the DK2 are all parallel, or come to a focus at the normal sample cell position, after which the rays would diverge. The latter is more likely, since they reported that the new light paths caused some loss of intensity, though the four extra mirrors would contribute to this. They compounded this problem by using square-section silica tubes for their cells, instead of cells with optically flat windows. They also supplemented their spectra with photoelectric microdensitometer measurements of photographic plates, exposed using a spectrograph. It would therefore be valuable if some of their results could be confirmed independently.

A furnace has been described[11] for use with a Cary 14H spectrophotometer based on the experience of a variety of furnace designs for use in molten salt spectroscopy. It has a stainless steel core and excellent temperature stability at temperatures up to 1050°C. It is shown in Figure 8. Here only the sample chamber

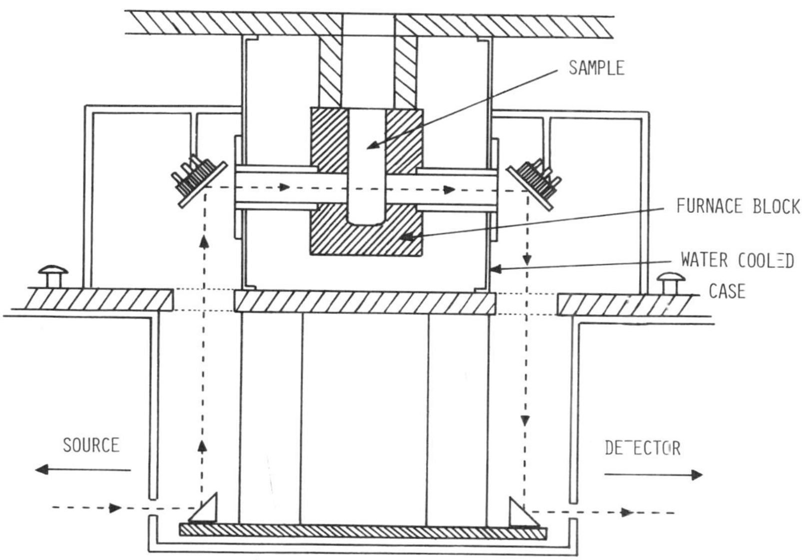

FIGURE 7. Modification of a Beckman DK 2 spectrophotometer for molten salt work.

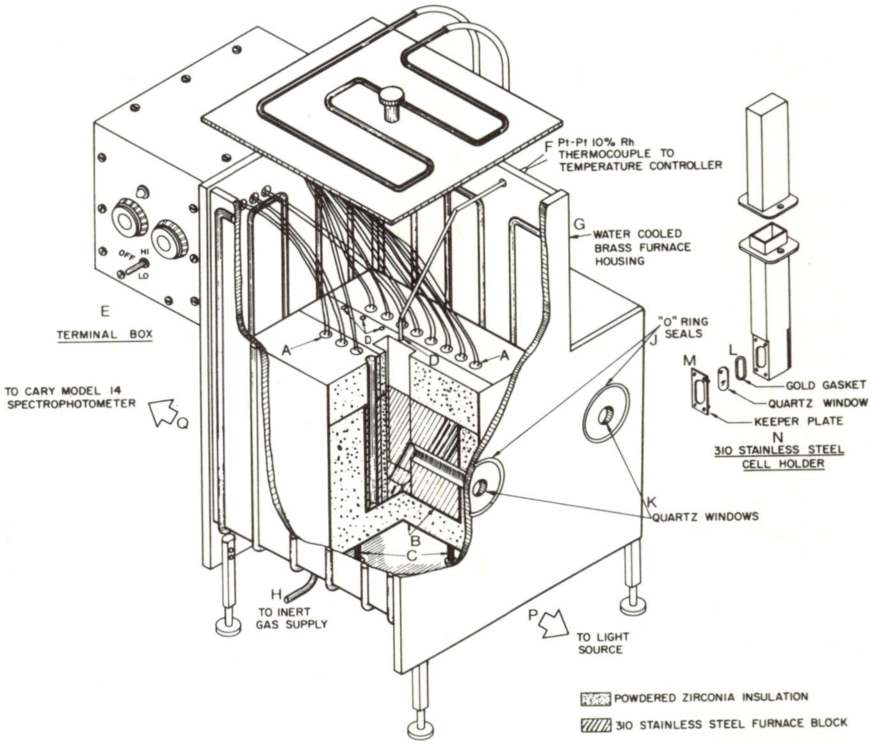

FIGURE 8. Exploded view of furnace assembly for absorption spectra measurements at temperatures up to 1000°C. A, 400 w firerod heaters; B, 310 stainless steel inner and outer walls of insulation container; C, positioning screws for furnace; D, thermocouple wells; E, terminal box; F, Pt–Pt—10% Rh thermocouple leads to temperature controller; G, water-cooled brass furnace housing; H, to supply of inert gas; J, O-ring seals; K, quartz windows; L, gold gasket; M, keeper plate; N, 310 stainless steel cell holder; P, to light source; Q, to Cary 14 spectrophotometer. Dotted regions represent powdered zirconia insulation and hatched regions the 310 stainless steel furnace block. Courtesy of The Chemical Society, London.

is maintained at high temperature and all spectra are recorded with air as reference. The method of heating was similar to that of Bailey and McIntyre,[40] using eight 400-watt Firerod heaters located in wells. Due to the excellent stability of the controller, coupled with the high heat capacity of the furnace, temperatures were constant to within ± 0.1°C for several hours over the entire temperature range; the thermal gradients in the light path were not reported.

To reach 1000°C and above, ceramic tube furnaces have been designed based on commercial tube furnace designs. The construction of one such furnace designed by Boston and Smith[12] and capable of reaching 1450°C is illustrated in Figure 9. Because the auxiliary heaters around the light ports were controlled independently from the ceramic tube heater, vertical thermal gradients were small. Simplifications[38] to this design evinced some difficulties.

Replacing the Pt-20% Rh alloy windings with Kanthal and without the optical port heaters, a vertical temperature gradient of 9.4°C cm^{-1} was found at 190°C, though the horizontal gradient was small, $<$0.2°C cm^{-1}. Under such conditions the temperature of the melt at the top of the spectral light beam could differ from that at the bottom by up to 13°C. With the guard heaters around the side arms in series with the main windings, the vertical thermal gradient was reduced to 5.5°C cm^{-1}. Part of the difficulty lay in the problems associated with winding the ceramic tubes. They are generally smooth and not helpful to the necessary closely spaced windings, particularly when attempting to have sufficient Kanthal heating wire located along the tube at right angles to the light beam. The ceramic cement which then must cover them to keep the windings

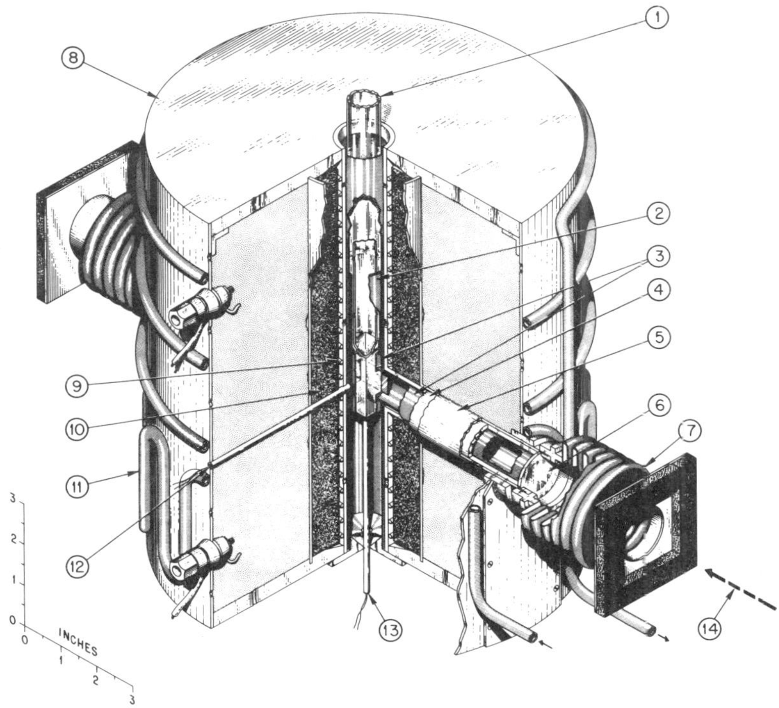

FIGURE 9. Furnace for use up to 1500°C in the Cary model 14H spectrophotometer. (1) Optical cell; (2) platinum cell holder; (3) platinum liner; (4) alumina liner; (5) metal shell; (6) silica window; (7) bellows; (8) end plate; (9) heater windings; (10) magnesia firebrick; (11) water-cooling coil; (12) control thermocouple; (13) sample thermocouple; (14) light beams; (15) closure block; and (16) diatomite powder. From *Rev. Sci. Instrum.*, Courtesy of the American Institute of Physics.

in place must be of the highest purity or the windings will soon burn out. When this happens, even if another wound ceramic tube is available the furnace would need recalibrating. Four Firerod-type heaters could be located around the inner ceramic tube, at the four corners of the sample cell, thereby not interfering with the light beam. The low conductivity of this type of furnace, however, may cause some difficulties in obtaining near isothermal conditions, and in regulation.

Another more successful tube furnace, designed for use at 1000°C but which operated successfully up to 1450°C, has been described by Boston and Smith.[37] It is also designed to have, like their 400°C aluminum block furnace, low thermal gradients, and was for use in their Cary 14H spectrophotometer. In some senses it is a hybrid furnace, having a metal core, (but not a solid block of metal) and ceramic tubes. The heater section consists of Inconel tubing, grooved and wound with sheathed cable. This overcomes the difficulty of attaching Kanthal wire precisely to ceramic tubing: the Inconel tubing fits into ceramic tubes. The inverted-T shape of the furnace is clearly seen in Figure 10. In order to reduce temperature gradients, the furnace windings were spaced more closely near the

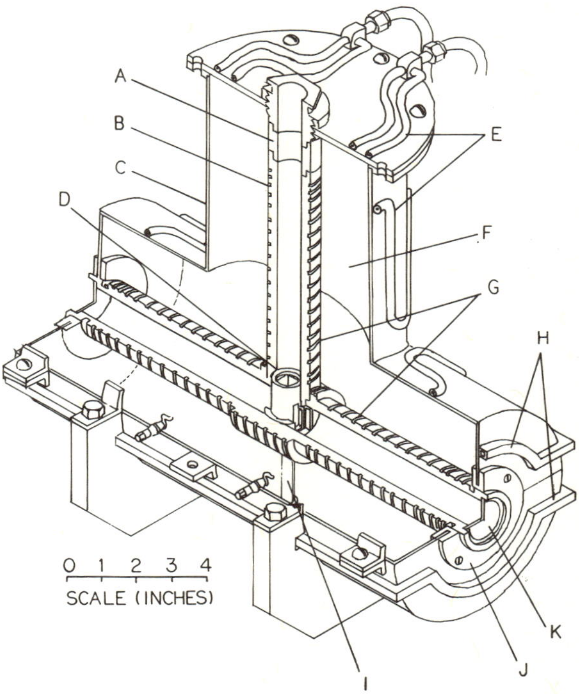

FIGURE 10. Cutaway diagram of the principal parts of the inverted T-shape 1000°C furnace. A, ceramic; B, sheathed heating element; C, outer jacket; D, cell holder; E, water cooling; F, insulated space; G, grooved Inconel tubing; H, water cooling; I, thermocouple well; J, Inconel collar; K, beam port window.

ends of the Inconel tubes than elsewhere; the windings were tapped to be able to vary the current through each section, using bypass resistors across the taps; and the vertical and horizontal elements operated off independent power supplies. Both were actuated by the same temperature controller. Maximum temperature differences of 3–4°C were obtained in the cell space, and ± 1°C in the window areas at 600–800°C. However, the vertical tube was open and when a guard-heater plug was used smaller gradients were obtained.

An alternative and novel approach has been that of Kerridge and co-workers[14] at the University of Southampton, England, who have utilized the local heating characteristics of radio-frequency (rf) heating. The apparatus is reported to have been working well for some time, but details have not yet been published.

4.2.2. *Inert Atmosphere Furnaces*

Although an inert atmosphere generally is maintained above a molten salt, specialized inert atmosphere furnaces are sometimes required. Many molten salts are corrosive, but the corrosive nature of molten fluorides is such that there is no suitable optically transparent container which they do not attack. Young has devised apparatus and techniques for handling these salts and measuring the absorption spectra of solute species in molten fluorides.

The furnace assembly is shown in Figure 11. Samples are supported in the intersection of an inverted T-shaped nickel tube, surrounded by a heated metal core. Various sized apertures are sited in the optical ports so that the detector only receives light that has passed through the sample. The T-section is vacuum tight and located between the silica windows in the ports and a lid on the top. The lid is compartmented and, once in place, the air in the furnace is replaced with helium. A thin plate separating the compartment in the lid and the furnace is removed (from between two O-rings) and the furnace and inside lid re-evacuated, flushed with helium, and pumping continued while the furnace is preheated to around 250°C. Then in a helium atmosphere the sample in its container is lowered into the light beam and furnace, and the furnace temperature raised sufficiently to melt the sample. The container and sample are essentially only in gas-phase contact with heated nickel, cooled silica windows, Viton O-rings, and Teflon gaskets.

4.2.3. *Thermal Gradient Furnaces*

A recent advance in molten salt chemistry is being brought about by the application of a thermal gradient furnace, this because the optics of a Cary 14H spectrophotometer have been shown to be capable of accepting two masks, each containing a narrow (*ca.* 1 mm high) horizontal slot (across the light beam), without significant loss in the performance of the instrument.[44,45] Soret coefficients of transition metal ions, and also the rate of transport of dissolved transition

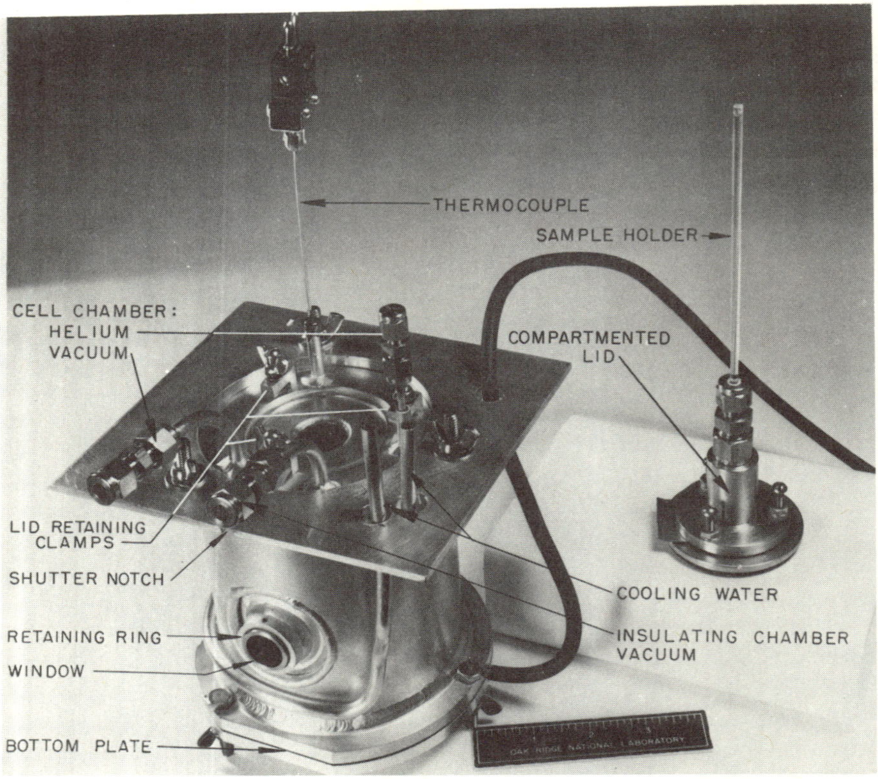

FIGURE 11. Furnace assembly used in the spectral study of molten fluoride salts. Courtesy of Marcel Dekker.

metal ions across a thermal gradient (leading to spectroscopically determined corrosion rates for metals by molten salts) are now being determined.[4,44,46]

The principle feature of this type of furnace (Figure 12) is that a pair of masks containing narrow, horizontal apertures remain fixed in the light beam, while the furnace is moved vertically to bring different sections of the melt into the light path created by the apertures of the masks.

The vertical section and horizontal light ports (V) of the furnace are made from aluminous porcelain tubing (i.d. 45 mm, o.d. 55 mm); the inner ends of the light port tubes were cut to fit the curved surface of the vertical tube. The Kanthal (12.6 $\Omega\ m^{-1}$) heating element (C) is a uniform coil around the outside of the vertical tube, with a total resistance of 100 Ω. The windings terminate just above the light port tubes, giving a thermal gradient of around 30°C cm^{-1} in the sample space. The furnace assembly rests in the sample compartment of the instrument on a brass plate (K), the furnace being held in position by three locating lugs (M), while the table is held rigidly and reproducibly in the leg clamps (L).

The inner mask (R) is an inconel screen containing an aperture 8 mm wide. It is mounted on a ceramic rod which is fixed to the furnace table by a locking

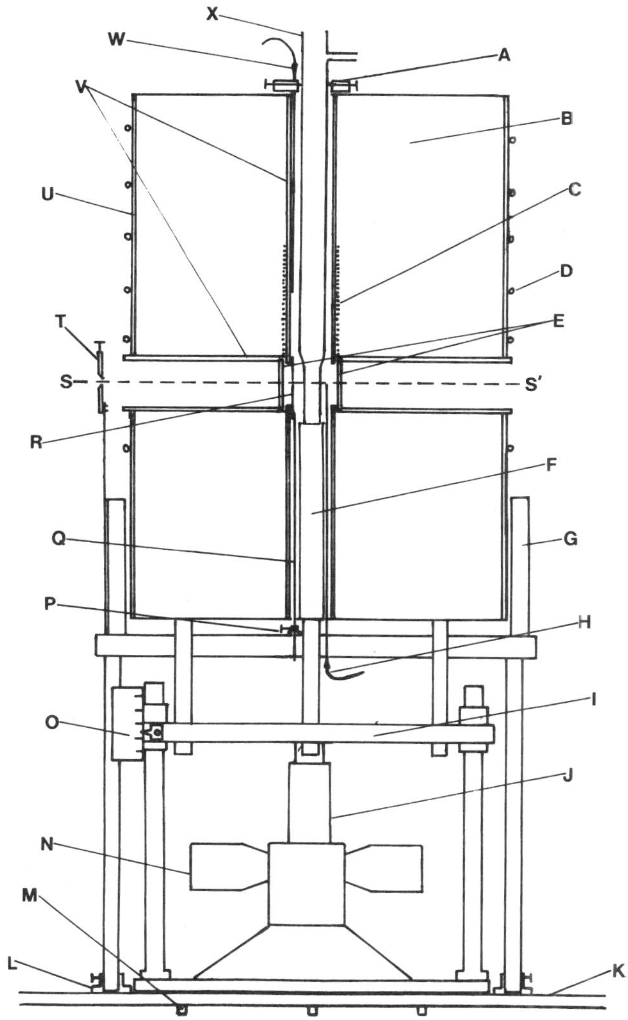

FIGURE 12. Thermal gradient furnace assembly, with silica optical cell in place. A, cell support collar; B, firebrick; C, heating coil (Kanthal); D, copper cooling tubes; E, silica windows; F, cell location platform; G, furnace support column; H, thermocouple; I, furnace support platform; J, 1 mm pitch screw thread; K, brass plate; L, table leg clamp; M, locating lug; N, graduated wheel; O, position scale; P, locking screw; Q, ceramic rod; R, inner mask; S-, S′, optical axis; T, outer mask; U, brass jacket; V, ceramic tube; W, control thermocouple; X, modified standard 10 mm silica cell.

screw (P). A second mask (T), also having an aperture 8 mm wide but of variable height, is mounted at the light exit port.

The thermocouple (Thermocoax, Phillips) (H) to measure the temperature of the horizontal section of melt being examined is also fixed to the table, with its tip in contact with the side of the cell, and in line with the apertures of the two masks. All three were adjusted to be in the vertical center of the light beam. The thermocouple was calibrated using a second, identical thermocouple contained in a tight-fitting silica sheath, and immersed in a melt contained in a cell.

Different horizontal sections of the melt are brought into the aperture by raising or lowering the furnace, using the graduated furnace position wheel (N). This controls the height of the platform (I) supporting the furnace *via* the 1 mm pitch screw thread. The furnace position, and hence the sample position, are known to ±0.005 mm, and can be altered by very small increments.

The maximum temperature at the top of the sample cell was 720°C, but the hottest portion of the vertical tube was around 900°C. A second more flexible furnace has been designed and built by Severn Scientific Limited, Thornbury, Bristol, England. Here there are two sets of independent windings, one above and one below the light beam area. The upper winding is of platinum, and therefore the top of the sample cell is capable of being maintained at 1000°C. The lower windings are of Kanthal. Consequently, the thermal gradient across a molten salt layer may be varied as required. At the time of writing this furnace has not been fully tested.

4.3. *Furnaces for Infrared and Raman Spectrophotometers*

Since the design of furnaces necessary for measuring the infrared and Raman spectra of molten salts is intimately bound up with the design of the container for the molten salt, a discussion of such furnaces is provided in the section on cell materials.

4.4. *Furnaces for Nuclear Magnetic Resonance Spectrometers*

Temperature control of samples in nuclear magnetic resonance and electron spin resonance spectrometers is a standard feature, but measurements at elevated temperatures are not. Rf heating of small samples can be employed with care. If heating wires are used they must preferably be nonmagnetic, and inductively wound. Platinum wire is commonly used.

The sample and the spectrometer rf coil, which needs to be a single coil,[47] may both be heated, but this limits the maximum temperature to the melting point of the silver wire used for the coil, 960°C. Alternatively, the sample may be heated and the surrounding coil maintained at room temperature. Here interruption of the heating current is necessary during the nmr measurement, and although higher temperatures can be reached, thermal gradients within the sample are likely. Hafner and Nachtrieb[47] have described two designs for single coil nmr probes and furnaces for molten salt applications. One furnace reaches nearly 1000°C, with the sample and coil being heated, and the other 720°C, with the coil at ambient temperatures. The coil consisted of 13 turns of silver wire wound internally in threads cut in a cylindrical lava shell. The heating element was noninductively wound, and consisted of 30 vertical loops of platinum wire set into grooves that had been cut in a surrounding thin wall lava cylinder. The whole was encased in a brass water-cooled jacket 40 mm in diameter. The salts were held in pyrex or vycor ampoules (volume *ca.* 1 cc), sealed under vacuum, and shaped to have an axial well for a thermocouple. A second probe containing the working reference was employed, and was readily and rapidly interchangeable in the magnetic field with the sample probe. It used a conventional marginal oscillator, and a separate frequency, but intercalibrated counter to record the resonance frequency.[48]

Using the furnace limited to 720°C, a small error arose due to the perturbation of the H_0 magnetic field by the dc current of the furnace coil[49] (around 5×10^{-5} in the chemical shift of ^{205}Tl). Only the molten salt resonances were affected, since the reference probe was held at room temperature. Because of the limited volume available for thermal insulation with this probe, and because the use of a metal core was precluded, thermal gradients of 5–10°C along the sample were experienced.[49] For ^{205}Tl studies this is not serious as $d\delta/dT$ is[47] around -0.5×10^{-6} deg^{-1}. Further details may be found in references 47–50.

Warren and co-workers have reported other designs and subsequent modifications.[51–53] Their first design[51] permitted nmr experiments between ambient temperature and 1300°C. It is easy to construct and has options for controlling the atmosphere around the sample and for applying uniaxial stress to the sample. Sufficient external cooling was found to be provided by the laboratory environment. To attain the maximum temperature of 1300°C the element windings were Kanthal Al, the thermocouples Pt–Pt + 10% Rh, and the receiver and transmitter coils platinum. Modifications which allowed a maximum of 1000°C were Chromel A-Tophet 30 windings, Chromel-Alumel thermocouples, and gold receiver and transmitter coils. Careful measurements of the magnetic field at the sample position showed that the combined effects of Chromel A heater elements and Chromel-Alumel thermocouples shifted the magnetic field by <0.1 G at 10 kG, provided the thermocouples were further than 6 mm from the center of the receiver coil. Typical temperature stability at 850°C was ± 0.05°C, with a maximum axial temperature variation over the central 30 mm of 5°C.

A more recent design[53] (Figure 13) has the sample at the bottom of an alumina tube, the other, room-temperature, end being connected to vacuum and pure argon sources. The closed sample end fitted into a platinum nmr coil wound

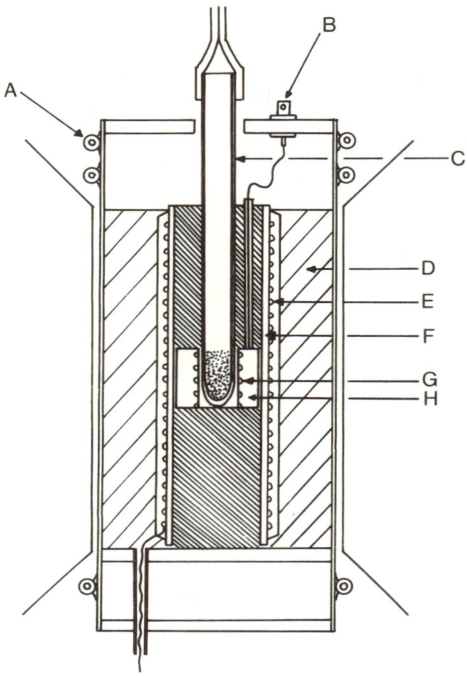

FIGURE 13. Schematic diagram of high-temperature NMR probe and sample arrangement. A, water-cooling coil; B, RF connector; C, alumina tube; D, fiberfrax insulation; E, platinum heater embedded in alumina cement; F, alumina furnace tube; G, platinum NMR coil; and H, castable ceramic form.

internally on a ceramic former. The platinum heating wire, on an alumina former, was wound noninductively and surrounded by thermal insulation. Water-cooling of the ends of the furnace assembly was employed. This design can be made around 60 mm wide. Warren and co-workers are currently using a design which allows ready clean-up should the molten salt escape from its container. This latest design[54] is 80 mm wide, and thus demands spectrometers where the pole gap can be adjusted. The heating wire is now molybdenum and the insulation zirconia fiber-board.

5. *Temperature Control and Measurement*

There are a large number of power regulators available on the market and improvements and modifications are continuously being made. It is therefore not appropriate to give an account of the various controllers that have been used over the years with the various furnaces just discussed, since many of the devices originally used are now no longer available. The reader is therefore advised to discuss his requirements with manufacturers and colleagues.

Some comments are appropriate, however. On-off controllers are to be avoided, as these tend to exhibit hunting and large overshoot characteristics; a proportional controller, with an anti-oscillation control is to be preferred. This author has found that S.C.R. (silicon controlled rectifier) controllers, with platinum resistance thermometers as the controlling thermocouple, provide stable and reproducible temperatures within his furnaces. The controller has operated well over many years, the only modification necessary being the insertion of a rapid electronic fuse, to protect the S.C.R. devices when furnaces occasionally shorted and burned out.

If water-cooling is employed, particularly if the top as well as the sides and bottom of the furnace are cooled, a safety system is advisable, such as a waterflow interlock system controlled by a solenoid valve which turns off the power should a leak develop anywhere in the system.

Thermocouples for temperature control and measurement should be chosen according to the range and maximum temperature up to which they are required to operate. For temperatures between 500 and 1500°C Pt–Pt + 10% Rh thermocouples are recommended. For temperatures from ambient to 1200°C metal-sheathed mineral-filled Chromel-Alumel thermocouples are recommended, preferably with 2 mm o.d. because of their ability to be bent easily yet retain precise location features.

The temperature of the molten salt inside a silica cell may be determined using a thermocouple in the melt at the center of the light beam. If the salt will attack the metal sheath the thermocouple may first be placed in a thin-wall silica tube. The original workers would calibrate this thermocouple against standard samples of tin, lead, zinc, aluminum, and copper metals and record its e.m.f. on an accurate potentiometer. Unless very precise temperature measurement is required, the present generation of digital display devices, with an accuracy of around ± 1°C, should be sufficient. However, a calibration graph of the readings of this thermocouple against the measuring thermocouple, located usually alongside the sample cell, should be constructed.

It is generally found that a good controller can maintain any temperature setting to within 0.1°, but the temperature of any point within the melt is constant to within ± 1°, unless, as described earlier, considerable care has been taken to reduce thermal gradients to a minimum.

6. High-Temperature Optical Cells

Gruen[11] has pointed out that there are four basic requirements for the containers of molten salts on which quantitative spectrophotometric measurements are required, *viz.:* (a) they must be transparent to electromagnetic radiation in the wavelength region in which measurements are to be made; (b) their softening or melting points must be considerably higher than the highest temperature at which measurements are planned; (c) they must be chemically inert to attack by the molten salt solutions; and (d) they must have a precisely defined optical path length, or one which can (readily) be determined by calibration. A fifth should be added, if good quantitative measurements are intended, namely that the two exterior and the two interior window surfaces should all be plain parallel and optically flat.

6.1. Materials Available

These conditions therefore limit our choice of materials. Optical quality fused silica is the best general material. In recent years, square-section transparent silica tubing has become available which is much cheaper than optical quality cells and particularly useful for preliminary experiments or where etching of silica cells is likely. These cells do, however, distort the light beam slightly, and if the position of the cell in the furnace is displaced, or the cell removed and not relocated precisely in the same position and orientation as before, absorbance values can vary slightly. Usually this is not more than a very few percent, but quantitative precision can be lost. Path lengths are also not truely obtainable. "Windowless" cells were initially developed for the particularly aggressive fluoride melts. Table I lists the main materials for uv-visible studies.

Standard 10 mm path length Pyrex or fused silica absorption cells may be used, and since they are nowadays constructed using a plasma torch, there are no cemented edges to fail at high temperature. Because the volume of powdered salt to be melted to fill the cell is two or three times the cell volume, it is

TABLE I. Materials for High Temperature Optical Cells

Material	Fused solvents	Upper temperature limit (°C)
Pyrex	Nitrates, some halides	500
Silica	Sulfates, chlorides, bromides, iodides, thiocyanates	1200
Alumina	Chlorides, chloride-fluoride mixtures, metal-molten salt solutions, glasses	1800
Magnesium oxide	As for alumina, alkali fluorides	2500[a]
Diamond	All solvents, including fluorides	800[b]
	Windowless cells	
Gold	Hydroxides	800
Platinum	Sulfates, all halides, oxides, carbonates, borates, silicates	1400

[a] Although this temperature is 300°C below the melting point of MgO, temperatures above 2000°C can only be attained in helium or argon atmospheres: hydrogen at these temperatures will reduce MgO.

[b] Above 600°C diamond begins to darken as it is slowly converted into graphite.

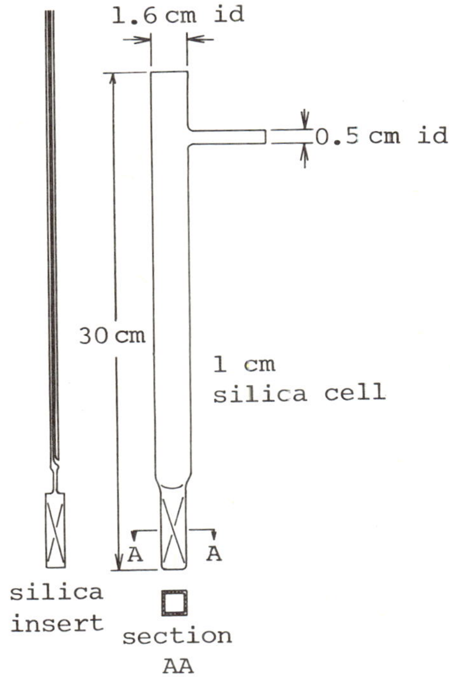

FIGURE 14. Optical silica cell and insert.

advisable to obtain cells with necks or to which necks can be attached. Figure 14 shows such a cell which has a fused silica spacer that can be used to stir the melt, reduce its path length as necessary, and provide the means for a suitable gas, usually inert and dry, to bubble slowly through or over the upper portion of the melt.

For the more aggressive melts, single crystal alumina and magnesium oxide have been shaped to have rectangular or square wells with plane parallel sides.[55,56] For liquid glasses at 1500°C, optically flat alumina windows have been cemented to a U-shaped section of alumina, and the molten glass dripped in.[57] After use the cell can sometimes be dismantled and used again.

6.2. *Windowless Cells for Corrosive Melts*

The small, but real inaccuracies attendant on the use of cells made with square-section silica tubing for weakly corrosive melts have already been mentioned. Windowless cells for the very corrosive melts such as hydroxides, oxides, carbonates, some silicates, fluorides, and metal-molten salt solutions have been developed, largely by Young and co-workers.[55,58,59] The original pendent drop technique, with its constant uncertainty of path length, was superseded by a windowless container called a captive liquid cell,[59] developed for use with wetting liquids (Figure 15). Molten salts are held at the level of a series of small holes, and the melt in the larger lower hole forms a double convex lens of reasonably reproducible path length which can be determined.[58,60] With nonwetting melts the small holes are no longer required. However, where possible, and particularly when accurate results are required, cells with windows are to be preferred. To this end the techniques used up to 1968 for uv-visible absorption spectroscopy of molten fluoride salts were reviewed and evaluated by Toth.[61] Numerous optical materials that could be useful for fluoride spectroscopy were suggested on the basis of thermodynamic predictions and limited testing. He concluded that quartz should be used to its practical limits and perhaps then should be coated with transparent metal films in order to extend its usefulness. Ultimately, however, diamond-windowed metal cells were viewed as the only corrosion-resistant absorption cell for fluorides above 500°C.

6.3. Designs and Window Materials for Corrosive Melts

Some of the limitations of windowless cells were early circumvented by Cocks, Schroeder, and Schwartz,[62] who designed the first fused-salt cell with diamond windows. This design is not readily suited to routine measurements.[63] Toth and co-workers[63] designed a cell with a graphite body that could be easily taken apart for cleaning. Although it was limited to systems inert to carbon, of which there are many such, the cell performed well. No fluoride melt leakage occurred around the windows, since most fluoride salts do not wet graphite. The maximum practical temperature is around 800°C, since above 600°C diamond begins to graphitize and above 800°C the rate of darkening is unacceptably fast.[64] Although this surface darkening can be polished away, prolonged exposures to these temperatures would necessitate the (expensive) replacement of the diamond windows and is thus to be avoided.

A possible future window material is boron nitride, a material which can withstand high temperatures and fluoride attack.[61,65] Pyrolytic boron nitride[66] is

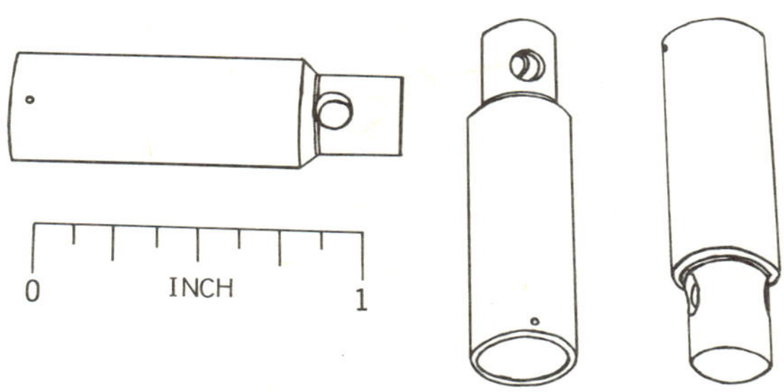

FIGURE 15. Captive liquid cells.

oxidation resistant up to 1300°C and has excellent thermal shock resistance. It is inert to most materials and not wetted by many. Also, it is unaffected by water, has good thermal conductivity, and maintains its strength up to 2200°C. (It should therefore be considered for furnaces where temperatures above 1500°C are required.) The transmission spectrum of a 1 μm thick boron nitride film, deposited on an alumina single crystal substrate, has been reported in the uv and visible region.[67] The spectrum is essentially transparent above about 500 nm and rises to an absorbance of unity around 400 nm. The etch-resistance of these films to chemical attack depended upon the chemical and the temperature employed, and on the deposition temperature of the film. Samples of a transparent form of boron nitride have been reported by Union Carbide. Whether it will prove resistant to attack by molten fluorides or be made commercially available for use as a window material remains to be seen. A study of the protection afforded against aggressive melts by thin boron nitride films on current materials is strongly recommended.

6.4. *Solving the Problems for Infrared Measurements*

Many infrared studies of molten salts have concentrated on the salts themselves because many ion fundamental vibrations have large intensities. Short path length cells are therefore required, particularly if the salts are also to be used as solvents. The corrosion of standard infrared cell windows by molten salts was first bypassed by Greenberg and Hallgren[23] who used a fine mesh platinum screen, another type of windowless cell, as a combined heater and support for molten salts. Mackenzie[17] has devised a wire mesh sample holder and controlled atmosphere furnace system (Figure 16). Hester and Krishnan[68] later extended this method, using stainless steel meshes etched in aqua regia to an optimum wire diameter. Even so, the smallest sample path length here obtainable is about one hundred times that necessary to record the most intense fundamental bands. Also, the path length cannot be reproduced in repeat experiments and fluctuations in sample refractive indices with wavelength tend to distort intense bands (medium intensity bands are affected less). This is a more serious problem than in visible region molten fluoride studies. Such devices do permit high sampling temperatures and enable the approximate relative intensities of weakly active vibrational modes to be recorded. Hester and Krishnan[16] have inferred the positions and structures of fundamental bands from observed overtone and combination modes. Griffiths and King[69] have recently compared these spectra with those of quenched melts recorded in KBr discs. They found general but not complete agreement with far infrared region data available from the effectively much shorter path lengths. It is therefore recommended that such studies of quenched melts can usefully be employed to supplement molten salt studies.

As indicated earlier, suitable window materials for infrared measurements are ultrapure silicon, the Irtan materials (particularly Irtan 4 and 6), and diamond.

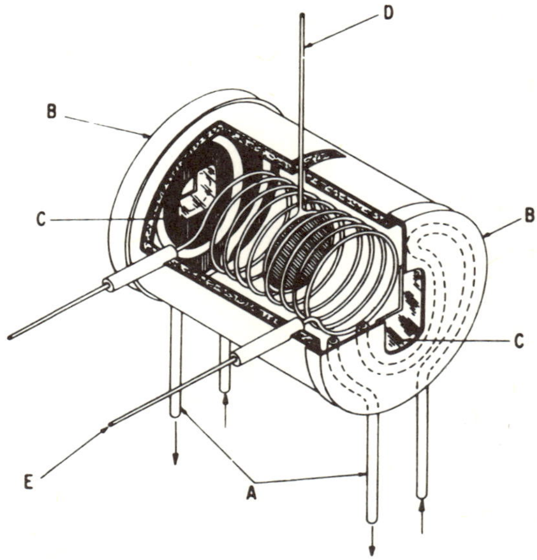

FIGURE 16. Wire mesh sample holder in controlled-atmosphere furnace. A, water-cooling coils; B, detachable end plates; C, MgO windows; D, platinum sample holder; E, Kanthal heater. Courtesy of Plenum Press.

Silicon can only be used below 400°C, at which temperature the thermal energy (*kT*) is that of the separation between the valence and conduction bands.[27] Also, due to high reflection losses the melt must wet the silicon surface. Acceptable transmission has been reported[15] between 200 and 4000 cm^{-1}, and for the Irtan materials which are more readily attacked than silicon. Diamond is limited to 600°C, but higher temperatures, up to a maximum of around 800°C, are possible for short periods (see above).

Some of the materials in Table 1 are not transparent to infrared radiation, and thus they have limited use. A possible future development, like that for uv-visible spectra, is the protection of window materials with a thin film of boron nitride. The near infrared transmission of boron nitride films at 500°C has been reported.[67] The transmission was generally above 60%. Further work in this area should be profitable.

Some details have already been discussed of the prism materials used in ATR spectroscopy in contact with molten salts.

6.5. Cells and Furnaces for Raman Studies

Cells and furnaces for use with Raman spectrometers which employ the Toronto arc mercury lamp are now obsolete and will not be described.

Since most continuous gas lasers have resonance lines in the visible region,

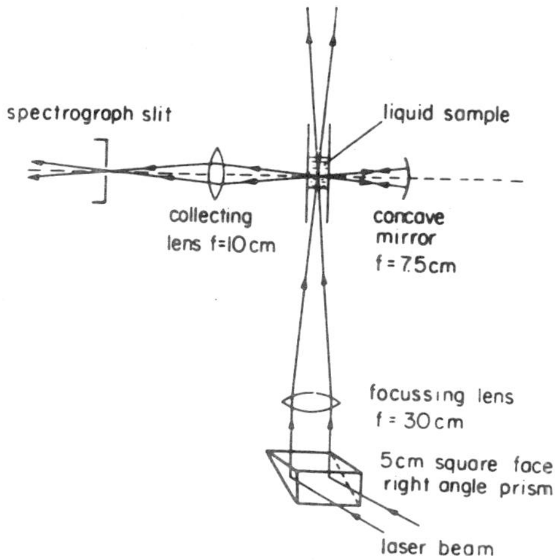

FIGURE 17. Optical system for laser excitation of molten salt Raman spectra. Courtesy of Plenum Press.

Pyrex and fused silica cells may be used. For aggressive melts a windowless cell, similar in concept to the captive liquid cells of Young,[59] was built by Solomons and co-workers.[70] The container, pyrolytic boron nitride, was used with molten fluoride up to 1030°C. (Whether there was any etching of the boron nitride was not reported—without specialized equipment it probably could not have been detected.) Another design by the same group,[71] using silica for the cell and a pulsed laser light source, could be used at temperatures up to 1000°C (Figure 17). Clarke and Hester[72] have simplified this design for use with continuous lasers (Figure 18). It features a fine porosity silica sinter, necessary to remove suspended particles, and does away with the laser beam entering or

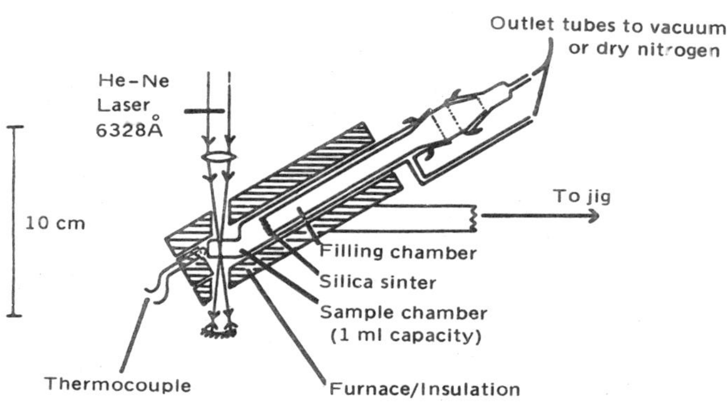

FIGURE 18. Molten salt Raman cell and furnace assembly. Courtesy of Plenum Press.

leaving through a liquid meniscus. The sample tube is silica and the furnace block (capable of reaching the softening point of silica, *ca*. 1200°C) is Syndanyo. It is wound with heating wire so as to minimize thermal gradients. Small holes in the block admit the laser beam and slightly larger cone-shaped holes at right angles to these allow the scattered light to exit. Focusing is achieved with a microscope objective lens, multipassing with a concave mirror below the cell and furnace assembly, and the back-scattered light reflected back through the sample by another concave mirror. A large aperture magnifying lens finally collects the scattered light and focuses it onto the entrance slit of a double monochromator.

7. Determination of Solute Concentrations

The Lambert-Beer law relates, for a fixed wavelength, absorbance to concentration and path length, *viz.:*

$$a = A/cb$$

where a is the absorbance, b the sample path length, and c the concentration. If b is in centimeters and c equals M, moles liter^{-1}, one can write:

$$A = \varepsilon bc$$

where ε is the molar absorbance.

To perform quantitative spectrophotometry A, b, and c must be accurately determined. In addition it must be shown that ε is independent of c over the concentration range studied. This latter condition regrettably is rarely tested systematically. A few results that agree within experimental error are often considered sufficient. Certainly, of the few studies that have been made to test Beer's law in a precise manner, all have found it obeyed accurately at all wavelengths. For example, Boston and Smith[73] demonstrated this with the visible spectrum of $NiCl_2$ measured in molten LiCl–KCl eutectic at seven concentrations, ranging from 0.01 to 0.4 M.

Absorbance values, whether measured directly using a suitable reference cell or by difference with air as reference are usually taken to be reliable. Nevertheless, the performance of the spectrophotometer should be checked regularly against a standard spectrum.

The tolerances of Pyrex and silica cells are given by the manufacturer, 10 mm cells being usually ± 0.02 mm. Any silica inserts used to reduce the path length can be measured with a micrometer screw gauge. Standard solutions of potassium chromate and dichromate can be purchased or made up, the molar absorbances being accurately known.[74] Path length changes due to cell expansion with temperature may be neglected, due to the very low thermal expansion of fused silica. Indeed, if it is necessary to quench a molten salt for subsequent analysis, the silica cell (Figure 14) at any temperature should be tilted so that the melt no longer completely covers the bottom of the cell and then plunged into cold water. The solidified salt will come cleanly away from the interior of the cell and can be readily tipped out as required. This cannot always be done with Pyrex cells. However, their thermal expansion over their more limited temperature range is no problem for accurate absorbance measurements.

Using the solidified salt, the concentration of the solute can be determined by many standard analytical techniques, though it is not uncommon for a component of the salt matrix to be an interferring ion. When a series of molten salt spectra are required and are most readily obtained by small additions of a salt, an alternative method for obtaining concentrations is preferred. If the density of the melt is known as a function of temperature, then, knowing the mass of salt weighed into the cell, the volume at any temperature can be calculated. The weight of solute added (after, say, the spectrum of the melt against air has been recorded) then enables its concentration to be obtained with sufficient accuracy, assuming concentrations of less than approximately 10^{-2} *M* are involved. This presupposes the addition has no effect upon the density of the solvent. If large additions are made one of two approximations may be used. It may be assumed that the added salt has the same density as the solvent. This is reasonable for chloride salts, added to alkali nitrate melts, where molar volumes are frequently additive.[75] Alternatively, if the density and temperature dependence of the solute are known, ideal behavior and mixing may be assumed. Discrepancies arising from the use of these approximations are expected to be less than those caused by completely neglecting volume changes resulting from the addition of solutes. Density data for many single molten salts have been collated by Janz and coworkers.[76] Little work has been done on mixtures. If it is thought necessary that such data is required for a system of interest, they may be obtained simply and accurately using Archimedes' principle and employing a platinum bob suspended on platinum wire. Experimental details have been published.[75,77]

If, as is common, a system is studied as a function of temperature, then because the solution expands and the path length of the cell does not, the concentration of the light absorbing species will appear to decrease with temperature increase, assuming the spectrum neither broadens nor shifts. A correction therefore should be applied, but this is rarely done in molten salt studies unless the spectra have been digitized (see later). Generally, again, the errors in omitting this correction are small. In one study,[78] all the (digitized) spectra were normalized to the volume obtaining at 300°C before any calculations or comparisons were performed. Such corrections should be made when they can be done routinely. The determination of molar absorbances of transition metal ions in chloride melts at temperatures up to 1000°C has been achieved with the same accuracy as for aqueous solutions at ambient temperatures.[11]

In the event that suitable window materials cannot be found, it is necessary to resort to windowless cells, and details of solute concentration measurements will be found in references 55, 58–61, and 79–81.

For infrared and Raman spectroscopy, concerned as they are largely with peak location and assignment, quantitative studies are rarely performed. Hopefully the advent of the new generation of ratio-recording infrared spectrophotometers will encourage such studies. To date, only a few quantitative studies by nmr have been attempted in molten salts.

8. Preparation and Purification of Salts Commonly Used as Solvents

Details of the purification of salts for molten salt studies will be found throughout this volume. For spectroscopic studies the main optical problems arise from decomposition of the melts and etching of the cell windows.

For fluorides, the detailed and scrupulous preparation, transfer, and handling procedures developed by Young[55,58,59,65] should be followed to obtain the minimum of corrosive attack, especially if seeking to test the suitability of boron nitride coated cells against fluoride attack. Really dry conditions to avoid etching of silica cells are necessary for alkali metal chlorides, particularly with melts above 500°C: bromides and iodides will exhibit some decomposition, with liberation of halogen at temperatures above about 600°C, if water is present, and will then etch silica cells.

Drying nitrates under vacuum at around 150°C for 48 h was found to be sufficient for most studies; the moisture content was then estimated at around 0.10%, the value commonly found after drying eutectic mixtures of nitrates by pre-electrolysis.[82] Using melt temperatures of below about 300°C a comparison of spectra, after several hours, of anhydrous and hydrated nickel chloride solutions in molten nitrate showed essentially identical spectra. It has been reported[83] that the slow bubbling of dry nitrogen through nitrate melts to which hydrated solutes have been added (up to around 10^{-2} *M*) will remove the water of crystallization if continued over a period of several hours. Thus nitrates do not have to be dried to ultra low levels of moisture content to obtain reproducible spectroscopic studies, unless measurements are required close to the decomposition temperature of nitrates containing transition metal salts, which is usually above 400°C.[82]

Molten sulfates, being acidic melts, do not generally attack silica cells, even at the high temperatures required to melt their various eutectic mixtures. Thermogravimetric analysis curves are often useful to determine the temperature range between which all the water of crystallization is lost and decomposition begins, so that salts may be dried with confidence.

In order to establish that materials and procedures are appropriate for the proposed experiments it is recommended that trial experiments be performed, using the non-optical square-section silica tubing in place of the much more expensive silica cell. Any unforeseen difficulties encountered can then or subsequently be overcome, and if at the end of the experiment the cell is not etched, the precautions taken will have been proved adequate. Further trial experiments should be considered to establish whether or not time-consuming or elaborate drying procedures are really necessary for the intended spectroscopic studies.

9. A Typical Experiment

A brief description is provided of the practical issues involved in performing a typical absorption spectroscopy experiment. Some hints and comments have been given already in the preceding sections; here it is hoped to include additional methodology. It is assumed that the reader is familiar with spectrophotometers and molten salt furnaces, and that the salts are prepared to the necessary purity and dryness levels.

9.1. Filling the Cell

The first task different from conventional procedures is filling the cell with the solvent salt. The exposure of the purified salt to the air must be kept to a minimum and the maximum use of dry-boxes is recommended. (For melts below 200°C and ones with organic, say quaternary ammonium, cations, these precautions are not so critical.) The clean, thoroughly dry cell is weighed, if possible in a dry-box. Unfortunately, static electricity can build up and affect the balance. The balance should be connected to ground and, in addition, a weak radioactive material placed in the balance: uranium dioxide is suitable. The salt, or eutectic, preferably powdered, but not too finely or transfer problems due to static electricity may be encountered, is then transferred to the cell, and the cell reweighed. Knowing the volume required, the working temperature, and the melt density, an approximate mass required in the cell should already have been determined. The cell is now sealed, usually with a bunsen valve on the side-arm (Figure 14) and a silicone rubber stopper. For high-temperature studies the cell may need to be transferred and attached to a vacuum line *via* a ground glass or silica joint, flamed-out, and sealed under vacuum. Normally a well-sealing stopper, through which the silica insert (Figure 14) can be raised and lowered, is sufficient.

9.2. The Insert

The insert (Figure 14) is connected to a gas supply so that dry nitrogen or some other suitable gas is passed slowly and continuously over the salt. The cell is then placed in the furnace. (It is generally important that the gas supply should be as dry as possible.) Once molten and when it is at thermal equilibrium, the spectrum of the melt in the cell against air under the various conditions or temperatures of the main experiment can be recorded. The insert may be used to stir the melt and help remove the occasional bubbles that arise on melting, but which are reluctant to rise to the surface due to the high viscosity of many melts. These bubbles interfere with absorbance measurements.

Note that the insert, if not being kept in the melt to achieve specific path lengths, will generally be at a slightly higher temperature than the melt, and can upset the thermal equilibrium of the system when used as a stirrer.

9.3. Visual Inspections and Avoiding Cell Failure

The cell may of course be removed from the furnace at any time for inspection, usually to check that no more bubbles are adhering to the internal walls, to see if all the solute has dissolved, or to observe the intromission and stirring characteristics of the insert. A (variable) flame or hot air blower (for low-temperature melts) should always be at hand, to prevent the melt solidifying. Should it start to solidify it must *not* be returned to the furnace, but heat must be immediately applied until it is properly molten again. If this is not done then the cell will continue to cool in the furnace, the melt will solidify and crack the cell, and the solid will then melt again, leaking out of the cell and into the furnace. Too much heating of the melt this way must also be avoided. The melt temperature is not known in these circumstances and the decomposition temperature of the melt may be exceeded all too easily.

9.4. Adding the Solute

Addition of the solute to the melt solvent can cause problems. A small stoppered bottle containing the solute is first weighed, using at least a five-place balance. An assistant is often useful at this point. The stopper of the cell is loosened and raised so that the small amount (a few milligrams) of solute may be tipped in from a spatula, the stopper is pushed home again, and the remaining solute in its bottle reweighed.

Various precautions must be taken. It must be ensured, by visual examination, that all of the solute has reached the melt, and that none is stuck to the sides or capillary rod of the insert. Tapping will generally dislodge such particles. If not, and provided the melt temperature is not too high, an alternative procedure is needed. The stopper and insert should be slowly withdrawn completely, care being taken to see that any drops of adhering melt have solidified before the insert is fully withdrawn. The solute will then generally fall into the cell without trouble, but before the spatula is used it should be touched against the cell mouth and the sample bottle several times to remove static electricity; otherwise a finely powdered sample will fly everywhere but into the cell. When the insert is returned to the cell, because it cannot be heated by flaming, it must not immediately be slid into the melt to act as a stirrer, or the melt will solidify. It is best to suspend the insert just above the melt for a little while, then remove the cell from the furnace and observe carefully, with a flame handy, what happens as the insert is moved slowly into the melt.

The sample weighing bottle should, when it is necessary to hold it, be held in cotton gloved hands with tongs to minimize weighing errors. Pre-weighed samples, which may be rapidly tipped into the cell, reducing possible moisture pickup, are to be preferred but are not always feasible (grounding of bottle and cell is still a prior necessity).

To avoid the static problems and particles which occasionally stick tenaciously to the neck of the cell the solute may be pelletized, in a dry-box, using a hand press, and the pellets broken down to lumps of approximately the mass required. These can generally be handled by tweezers, and dropped precisely down the middle of the neck of the cell: tipping from a bottle usually produces sufficient sideways motion to cause the lump to hit and perhaps stick to the inside of the neck above the melt. Particles stuck to the sides, if not too high up, can be captured by the melt, by tipping the cell until the melt reaches them; particles on the stem of the insert cannot be recovered in this way.

9.5. *Aiding Dissolution*

The introduction of one or two lumps of solute into the melt can introduce other problems. The small surface area means that the solute dissolves very slowly indeed, because the solute frequently is not wetted by the solvent. The insert can sometimes be used to break up the lump and agitate the solution, but dissolution can still take several hours, during which time the spectrophotometer can drift from the settings used to measure the reference spectra, or the melt may begin to decompose or etch the cell. Thus the problems of powders may have to be accepted, particularly if a series of spectra is required as a function of concentration. It is sometimes possible, however, to prepare a melt of solvent containing a large concentration of the solute. This may be quenched, a sample analyzed, broken into pieces, and a medium-sized piece added to the melt. This will dissolve more quickly, but the increase in volume of the solvent must be recognized.

When dissolution is very slow, a wire stirrer of platinum or suitable material can be introduced into the melt and made to move continuously by a simple cam arrangement. Alternatively, the cell may be placed in an inclined rocking furnace, and a cam adjusted so that the molten salt flows slowly from the cell and partway into the neck and back again. The furnace may be simply constructed from a transparent silica tube wound with heating wire, but with sufficient space between the coils so that the movement of the molten salt may be observed. A thermocouple alongside the cell is used to control and maintain the furnace at the correct temperature. The author has sometimes had to have melts stirred or rocked continuously for two days to get some solutes to dissolve. Anodic dissolution has been used at times, but difficulties have been experienced, particularly with the platinum group metals. The temptation to speed dissolution by increasing the temperature should be avoided—decomposition of melts, particularly nitrates, often occurs, together with etching of the cell.

It should be noted that some of the previously described furnace designs may not be readily adaptable to allow multiple additions of solute to the cells they perforce use.

10. Future Work, Equipment and Techniques

Absorption spectra studies have been made for most of the metallic elements in the periodic table. For many only a survey has been performed in the most common melts, usually chlorides. Present and future studies require a more detailed examination of such spectra, particularly in fused salts containing various ratios of two different anions. Such studies, for speed and efficiency, require the ability to record a series of spectra, on varying the melt characteristics, within one basic experiment. Flexibility should be of high priority in choosing specialized equipment design.

10.1. Modern Treatment of Results

Modern spectrophotometers are now generally supplied with microprocessors, or are computer-compatible. Indeed, it was one of the pioneers of molten salt spectroscopy, G. P. Smith, of the Oak Ridge National Laboratory, Tennessee, who, besides initiating the Cary 14H series spectrophotometers for measuring molten salt spectra, also organized the Instrument Division of ORNL to build the world's first direct digitizing device for spectrophotometers, the transistor having just been invented. The original purpose was to permit the ready and accurate subtraction of the reference spectrum, recorded against air, from the sample spectrum, similarly recorded. Other manipulations of spectra were soon performed (beyond the simple procedures of changing ordinate and abscissa axes, scaling, and presentation as graphical output). For example, Brynestad and Smith[84] investigated mathematically the generally accepted concept that the presence of one or more isosbestic points in a series of spectra meant that only two species were present in equilibrium. They extended this concept and defined its limitations. They consequently utilized their digitized spectra to compute, for the first time, the spectra of species that cannot be measured free from overlapping absorptions due to other species which are always present in solution.

10.2. Digitizing Spectra

Since computer-compatible spectrophotometers may not be available to everyone, various methods of enabling instruments to output digitized spectra are now briefly described.

10.2.1. Principles of the Digitization Process

Digitization may be described as the process by which a physically continuous function, such as an absorption spectrum, is transferred into an array of numbers which accurately represents the original function. A spectrum is most simply digitized by recording both the absorbance and the wavelength (or wavenumber as appropriate) at suitable, though not necessarily equal, intervals of wavelength. By this method, each digitized point is accurately defined on both the x and y axes. The digitizing process may be related to the gradient of the spectrum—the larger the gradient, the closer the wavelength interval of digitization.

The more useful procedure is to digitize a spectrum at equal intervals of wavelength. The bulk of the data then required to represent a spectrum is approximately halved, as the wavelength corresponding to every digitized absorbance need not be stored. If (a), the initial wavelength of digitization, (b) the wavelength interval, and (c) the total number of digitized absorbances are known, the wavelength corresponding to any sequentially digitized absorbance can be calculated. Perhaps the most important advantage of digitizing spectra at equal wavelength intervals is the simplicity with which a variety of mathematical procedures (such as smoothing or differentiating spectra) may be applied to such data.

10.2.2. Adapting Spectrophotometers to Produce Digital Spectra

Digitizing techniques have been applied to a wide range of spectroscopic methods. Lipsett and co-workers[85] have described a system digitizing fluorescence and excitation spectra, fitting their spectrometer with a wavelength shaft encoder coupled directly to the monochromator drive shaft. This encoder made 1000 counts per 5 nm of scanned wavelength. The photomultiplier current was divided into 500 distinguishable levels by an analog-to-digital (A-D) converter, the signal being smoothed electronically. Digitizing intervals of 0.005 to 0.160 nm were employed and the monochromator shaft position and spectral intensity were both written directly onto magnetic tape. Subsequent computation was required to convert this data into wavelength and absorbance values respectively.

A system for automatically digitizing infrared spectra has been described by Jones.[86] Standard scan speeds of 40 cm^{-1} per min and encoding intervals of

0.5 cm^{-1} were chosen. A complete spectrum (4000-280 cm^{-1}) took $3\frac{1}{2}$ h to record, including the measurement of four calibration spectra. Information was stored on punched tape. Modern infrared spectrophotometers are not slowed down by the digitizing process. Digital methods in Raman spectroscopy have been described by Janz and Downey.[87]

Butler and Hopkins[88] employed a small computer coupled directly to a uv-visible spectrophotometer *via* an A-D converter. One channel was connected to the photometer with a digitized resolution of 0.0006 absorbance units. The other scanned the output of a precision potentiometer geared to the monochromator drive shaft (digitized increment, 0.125 nm). The real-time average of 128 absorbance signals was computed and stored for each digitized wavelength during a spectral scan.

Sakalys[89] has developed a similar real-time signal averaging technique using a dual-scanning spectrophotometer coupled directly to a programmable calculator. Wavelength drive was achieved through a bidirectional stepping motor controlled by the calculator to within a positional accuracy of ± 0.08 nm.

In the author's group similar, but simpler, equipment is used. The system comprises two digitizers, five logic modules, and a tape punch unit (Figure 19). The absorbance reading is encoded *via* a friction pulley on the pen drive wire. The pulley drives a single turn digitizer, type 12/ST/1000, encoding 0 to 999 in

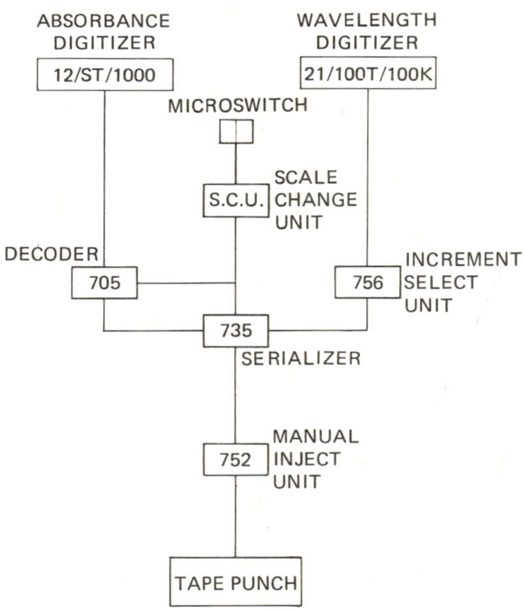

FIGURE 19. Digitizing system fitted to Cary 14H spectrophotometer.

one turn. A microswitch controls the scale change unit (SCU) logic module, which generates 0 or 1 for the 0–1 absorbance scale, and 1 or 2 for the 1–2 scale. The wavelength position is digitized by a multiturn digitizer, type 12/100T/100K, giving 100 turns, and 1000 parts per turn. It is connected to the monochromator drive shaft through antibacklash gears, and encodes from 0 to 30,000 at 200 parts per turn over 150 turns (200 nm per turn).

The remaining four logic modules operate as follows. A model 705 three decade decoder gives a visual display of the pen position from the absorbance digitizer and a parallel output to the serializer. The number generated by the SCU is also displayed and output to the serializer. The increment selector, model 756, determines the sampling interval by accepting the output from the wavelength digitizer and provides pulses at intervals of 0.1, 0.2, 0.5, 1, 2, 5, or 10 nm, as chosen by the operator. The serializer, model 735, accepts the output from the decoder (705) and the SCU, and feeds it, in serial form, to the tape punch unit, *via* the manual inject unit, model 752. The serialization cycle may be initiated manually, or automatically by the increment selector module (756). The manual inject unit (752) enables single digits of information to be punched onto the tape for identification purposes. The tape punch unit is a standard Facit-Ohdner model 4070, having a maximum speed of 70 characters per second.

To date there are very many reports of different ways to make spectrophotometers computer-compatible. Since the devices employed in this particular technology are now developing so rapidly in sophistication the reader is advised to consult an electronics specialist in the present state of the art as to the best and most economical way to obtain digitized spectra. At the time of writing the current trend appears to be the replacement of the wavelength drive with a stepper motor, or the incorporation of an absolute optical encoder, and the direct recording of the pen signal onto magnetic tape. Even if a suitable tape punch unit and other equipment are available, it may be cheaper, and it is obviously more efficient (assuming any data signal errors are detected and eliminated automatically), to avoid the electrical-to-mechanical-to-electrical sequence.

10.3. Mathematical Smoothing of Spectra

Noise is inherent in all spectroscopic measurements and, as indicated earlier, molten salt spectra generally contain more noise than room temperature spectra because of the heating effect on the detector of the thermal emission of the molten salt. Normally, infrared and Raman spectra are intrinsically much more noisy than uv-visible spectra.

Traditionally the way to reduce noise has been to average many recordings of the same event, as noise signals are random positive and negative contributions to the signal and average to zero, whilst the information present is always positive. The average of a large number of absorbance signals at each digitized wavelength in a spectrum would be found. Unfortunately, the improvement in signal-to-noise varies as the square root of the number of measurements and, for example, a twentyfold improvement is very time-consuming (and assumes drift-free electronics in spectrophotometers). Such an improvement has been achieved, for the same time expended, in nmr studies, by changing from continuous wave to Fourier transform measurements. For molten salts there is the added problem that the cell may be etched, or the melt slowly decomposed, during the multiple recording process.

Mathematical procedures which reduce noise can be applied to digitized spectra. Savitzky and Golay[90] have discussed the smoothing of spectral data by a simple moving average and with nonlinear triangular and exponential smoothing functions. They discounted such methods since they lead to distortion and reduction in intensity of the peaks. Instead, Savitzky and Golay evaluated an exact least-squares method, from a procedure derived theoretically by Worthing and Geffner.[91] It is simple to implement and involves multiplying a sequential set of 5, 7, or 9 . . . data points by a corresponding number of 'convolution coefficients'. The sum of this set is divided by a normalizing factor. The resultant is a least-squares fitted, smoothed value of the center data point of the set. The data set is then moved forward by one point, and the process repeated until the whole spectrum is smoothed. If the set contains 9 data points the operation is referred to as a 9 point convolute. Essentially the process involves the fitting of a polynomial, of degree n ($n = 2$ to 6) to an array of data points recorded at equal sampling intervals. The values of the convolution coefficients are determined by the order of the polynomial chosen.

This convolution method has the advantage that it is noniterative, and therefore requires very little computing time. The degree of noise reduction is roughly proportional to the square root of the number of points used in the convolution. Spectra characterized by insufficient data points may suffer degradation as a result of this smoothing procedure. This is most common when large digitizing intervals are used to record narrow absorption bands, because the convolution procedure then sees these peaks as catastrophic noise spikes and attempts to reduce their amplitude accordingly. Additionally, spectra will be degraded if the number of points used in the convolute is too large.

The tables of convolution coefficients given by Savitzky and Golay[90] contain a number of mistakes, resulting from an error in their reasoning. This was noted by Steinier and co-workers,[92] who published the corrected coefficients.

Modern spectrophotometers, fitted with a microprocessor and having a relatively small memory, are potentially capable of smoothing their spectra by the above procedure, but this is rarely done: instead they differentiate their spectra.

10.4. Differentiation of Spectra

Figure 20 shows the first four derivatives of a simulated absorption profile: the ordinate scales are not comparable. Odd-order derivatives are zero at the (observed) peak maximum, and therefore can aid in pinpointing an (observed) peak maximum. Should the profile contain additional bands odd-order derivatives will identify where the observed peak maxima are, but not necessarily the true peak maxima, particularly if there is significant overlap of bands. The second and fourth derivatives are similar in shape (the second derivative being inverted) to the original band, but are much narrower. In the case of Gaussian bands, the second and fourth derivatives have half-widths (i.e., the full width at half-height) which are 53 and 41%, respectively, of the original band half-width. For a Lorentzian band, the corresponding values are 33 and 20%.

An even-order differentiated spectrum therefore enables a clearer picture to emerge of the correct number and position of the bands within an absorption

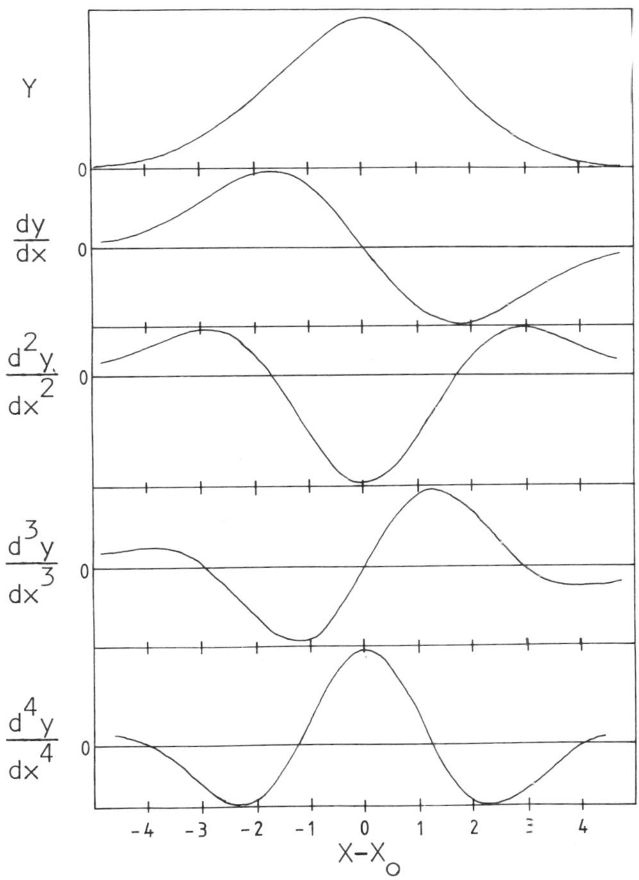

FIGURE 20. Computer simulated absorption profile and its first four derivatives. Ordinates are not to scale: second and fourth derivative minima and maxima are a factor of around 10^{-2} and 10^{-3}, respectively, smaller than the original maximum.

profile. The peak maxima identified thereby will now be much closer to the true maxima.

Some modern automatic spectrophotometers will draw the first, and sometimes the second, derivative of a profile, using a mechanical device which requires a constant scanning speed. The heights of differentiated spectra are much less than the zeroth order spectrum. If the recorder absorbance scale is increased to compensate for this, the noise in the spectra is increased by the same amount, generally resulting in a very noisy spectrum. This spectrum is often not particularly informative. Since these spectrophotometers have no means of smoothing their derivative or original spectra no improvement in spectra is directly attainable.

Spectrophotometers with microprocessors often contain programs which will calculate derivative spectra. The convolution method of Savitzky and Golay[90] is generally used, since it can calculate the derivatives of a spectrum by using the appropriate set of coefficients. Again, difficulties can arise.

Firstly, the zeroth order spectrum is not smoothed before it is differentiated, and the resulting spectrum is generally quite noisy. A fourth derivative spectrum can often, when expanded to fill the width of the chart, consist of rapid oscillations, which are noise spikes. These may be as much as twenty times the magnitude of the fourth derivative peak heights. This was observed on a Perkin Elmer 557. The latest Gilford spectrophotometer, model 2600, records spectra at much closer wavelength intervals, and the absorbance at each wavelength can be the mean of a number of determinations. This effectively smooths the original data, and hence the derivative spectra it subsequently calculates are little troubled by noise. The Perkin Elmer instrument could probably achieve this performance if the original data were smoothed.

A possible problem with these manufacturers' approach is that the number of convolution coefficients is either fixed or very limited, a 9 point convolution seemingly being favored.[93] The wavelength spread of the convolution process may be altered, i.e., the convolution coefficients may operate on adjacent recorded data, or every second, or third data point, etc. This is not helpful and a consistent pattern is not always detectable between spectra as this parameter is varied.

Although the convolution coefficients do smooth spectra to a certain extent in the differentiation process, the amount of smoothing is dependent upon the size of the convolute. A 9 point convolute will only improve the signal-to-noise ratio by a factor of about two. We have digitized spectra at small wavelength intervals and have been able to employ up to 49 point smoothing and differentiating convolutes.[3,44,45] The resulting profiles were essentially free from noise.

10.4.1. Second and Fourth Differential Spectra

An important feature of the second and fourth differential spectra of a profile containing several bands is that the largest and broadest band essentially disappears, because its height is now considerably less (often two orders of magnitude) than the derivative heights of smaller and narrower bands in the original profile. The mathematical aspects will not be here discussed, but it can readily be shown that for a system containing two nonequivalent Gaussian bands, the second and fourth derivatives have minima and maxima which are inversely dependent upon the half-widths to the power two or four, respectively. For example, in the second derivative, for two bands of the same height, but with one equal to half the half-width of the other, the ratio of peak heights will be 4 : 1 in favor of the narrower band. For the fourth derivative, this ratio becomes 16 : 1. As an empirical rule the use of the fourth derivative for the analysis of overlapping spectra will highlight smaller bands, while virtually ignoring the broad, main bands. Conversely, the second derivative tends to show both broad and narrow bands, but of course has poorer resolution capabilities than the fourth derivative.

To minimize the possibility of failing to locate bands during analyses and of subsequently making spurious assignments, it is recommended that both second and fourth derivative spectra be calculated and then compared. Good agreement between the peak positions obtained separately will indicate that the band in question is neither spurious nor far from its true position.

We have found[3,44,45] that for transition metal spectra, charge-transfer bands essentially disappear upon even-order differentiation. Spin-allowed transitions become clearly identified, and spin-forbidden bands appear with approximately the same heights as the spin-allowed bands.

10.4.2. A Powerful Example: Resolving the Spectra of Iron(III) Complexes

To show the power and potential of this approach, which will permit in future a significant extension in the information and assignments possible for molten salt spectra, an example is now outlined.

The spectrum of d^5 iron(III) ion in molten (Na,K,Al)SO_4 (cation mole ratio 17 : 17 : 6) is dominated by a strong ligand-to-metal charge-transfer absorption edge, which completely swamps the weak, spin-forbidden *d-d* transitions. The effect of chloride addition shifts this edge progressively to lower energy (Figure 21), as does temperature increase. Weak bands are detectable on the rising edge, but are never of sufficient intensity to form a well-defined shoulder. With the complete replacement of sulfate in the melt by chloride the edge is shifted further into the visible region, and the weak bands evident at lower chloride mole fractions are no longer apparent.

Second and fourth derivative spectra were calculated, using 35 point derivatives of 35 point smoothed spectra, with a polynomial of order five. A large number of weak, closely spaced bands were identified. The variation of these band positions with chloride content was small, and temperature-independent. A representative plot is shown in Figure 22 and typical second and fourth de-

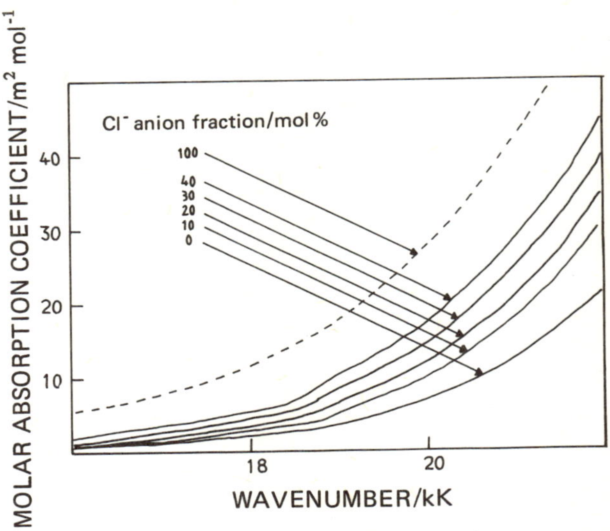

FIGURE 21. The effect of chloride addition on the iron(III) spectrum in (Na,K,Al)SO_4 at 699°C. Dashed line, (Na,K,Al)Cl melt. All spectra have been subjected to a 47 point smoothing, fifth order polynomial.

rivative spectra in Figure 23. Since chloride and sulfate have adjacent positions in the spectrochemical series for molten salts,[94] and the energies of the resolved bands are virtually independent of chloride concentration, it may be concluded that no coordination change takes place with increase in chloride content. Electrostatic factors will favor fourfold coordination, so tetrahedral complexes of iron(III) will be present.

The observed bands may be assigned with reference to the spectrum of tetrahedral $FeCl_4^-$ in acetone at room temperature,[95] and in the solid pyridinium salt $(PyH)_3Fe_2Cl_9$ at liquid hydrogen temperature.[96] These bands (shown in Table II), were identified by eye and the assignments for the $FeCl_4^-$ spectrum are due to Balt.[95]

It is to be expected that the $FeCl_4^-$ spectrum measured at 20 K would reveal more bands than the number seen at ambient temperatures. The comparison provided in Table II is with the bands (recorded nearly 1000°C higher) and identified by derivative analysis of molten salt spectra. The low temperature bands were found, and more. In addition, most of the band maxima identified in molten (Na,K,Al)SO_4 were also at slightly lower energy, that is, shifted in the direction expected with temperature increase. Further, comparison of the peak maxima identified in the sulfate and chloride melts showed that, for tetrahedral complexes of iron(III), sulfate and chloride have identical, rather than adjacent,[94] positions in the spectrochemical series for molten salts.

It may therefore be concluded that, in the case of iron(III), differentiation of observed spectra is a powerful and sufficiently precise technique for identifying spin-forbidden bands, even when those bands are broadened by high temperatures. Differentiation of the spectrum of $FeCl_4^-$ at 20 K should therefore reveal the additional bands found in the melt spectra, plus possibly some more. The additional bands can be identified with sextet–doublet transitions, or splittings of quartet term states: those possibly to be identified from low temperature spectra will no doubt assist in better assignments.

10.5. Band Fitting

It has long been the aim of spectroscopists to resolve, using a computer, observed spectra into their overlapping Gaussian or Lorentzian bands. This in-

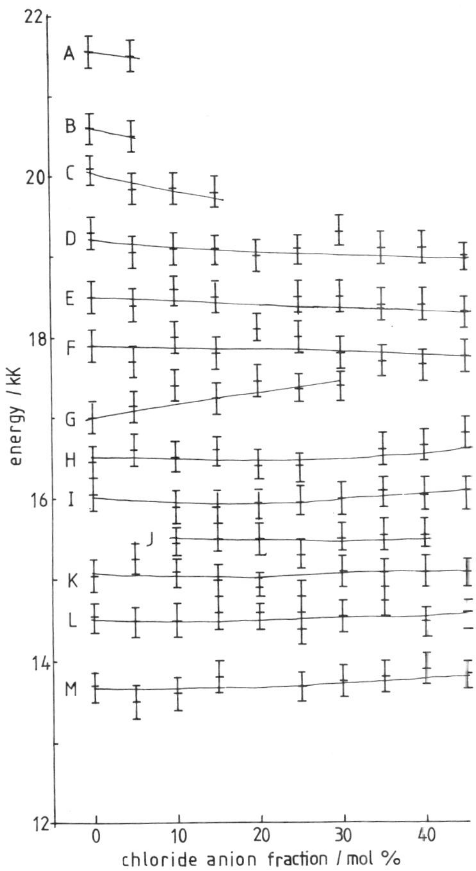

FIGURE 22. Resolved band maxima from iron(III) spectra in (Na,K,Al) (SO_4,Cl) at 610°C. Band assignments given in Table II.

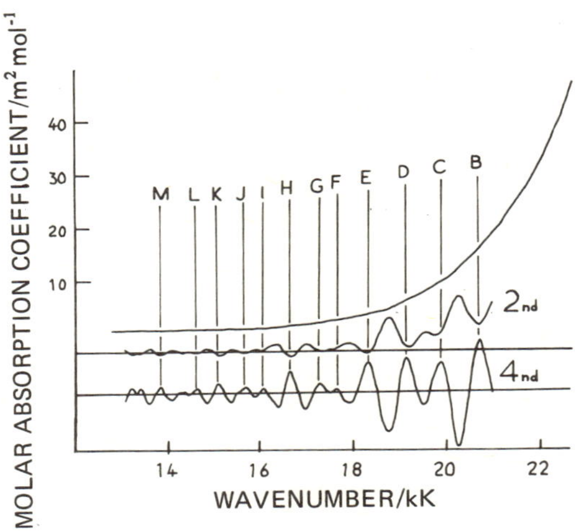

FIGURE 23. Spectrum of iron(III) in (Na,K,Al)(SO_4,Cl) containing 20 mol anion % chloride and its second and fourth derivative spectra. Band assignments given in Table II.

deed is often reported, but the levels of success vary. For example the uniqueness of the fit may often be questioned. 'Better' fits may be obtained if additional bands are proposed within the envelope. A serious study has not really been made between the need for these additional bands and the improvement in fit obtainable if the equation describing the bands is modified. For instance, infrared

TABLE II. Comparison of Maxima (cm^{-1}) of Resolved Bands from Iron(III) Spectrum in (Na,K,Al)SO_4 and (Na,K,Al)Cl Melts with the Reported Maxima of $FeCl_4^-$

Band	*Assignment*	*$FeCl_4^-$ in* acetone at 300 K	*$FeCl_4^-$ in* $(PyH)_3Fe_2Cl_9$ at 20 K	*Fe(III) in (Na,K,Al)Cl at 972 K*	*Fe(III) in (Na,K,Al)SO_4 at 1003 K*
A	$^4T_1(^4P)$	22,000[a]	22,100	nf[b]	21,500
B		nf	nf	nf	20,900
C		nf	nf	nf	20,000
D	$^2\Gamma$	nf	19,750	nf	19,500
E	$^4E(^4D)$	18,760	18,800	18,800	18,700
F		nf	nf	18,300	18,300
G	$^2\Gamma$	nf	17,800	17,700	17,600
H	$^4T_2(^4D)$	16,700	16,500	16,900	16,700
I	$^4A_1(^4G)$	16,150	16,150	15,900	16,200
J	$^4E(^4G)$	nf	15,650	nf	15,600[c]
K		nf	nf	15,100	15,100
L	$^4T_2(^4G)$	14,600	14,600	nf	14,400
M	$^4T_1(^4G)$	13,700	13,700	13,200	13,400

[a] 883 K.
[b] nf = not found.
[c] 943 K

spectra fit best to a Lorentzian band shape and uv-visible spectra, particularly around band maxima, fit best to a Gaussian shape. The observed wings of such spectra are generally broader than required by Gaussian bands: Lorentzian bands have much wider wings. The suggestion is at times made, because the sum of the individual Gaussian bands does not match properly the observed profile in, say, a certain wavelength range, that a small additional (Gaussian) band is needed in this region to complete a 'proper' resolution. This overlooks the genuine possibility that if the computer program were modified so that each Gaussian band could have a certain percentage of Lorentzian character, the need for the proposed extra band might disappear.

Previously the number of bands, and hence transitions, associated with a profile has been determined by the empirical principle that a profile should be resolved into the minimum number of bands consistent with an acceptable chemical explanation: the current work and possibilities available using even-order derivative spectra now permit a much better basis for stating the number of bands within a profile. Derivative spectra have the advantage over curve-fitting methods in that no assumptions have to be made as to whether the bands are Gaussian, Lorentzian, log-normal, etc.; no estimates are needed of their heights, half-widths, or positions, or even of the number of component bands. Indeed, derivative spectra can yield reasonable estimates of the band parameters.[97]

Two recent reviews on band-fitting, by Maddams,[98] and by Barker and Fox,[99] deal adequately with the problem, but the most recent advances of smoothing and differentiating spectra to determine the number of bands present were not available to them. They, particularly the former, do however stress the difficulties and pitfalls present in seeking to fit several overlapping bands such that a good match is obtained with the recorded profile. They certainly do not have any procedures which can be used to resolve the iron(III) spectra we have discussed above (Figures 21 and 23).

Previous efforts in curve-fitting have been largely for the purposes of identifying only the number of bands and, hence, the species present in a system; or where the band maxima are located and how they change when a constraint is applied to the system. It is proposed here that this information is much more readily available (without the need for large computers) by proper smoothing and differentiation of spectra rather than any curve-fitting technique.

10.6. Other Applications of Digitized Spectra

Mention has already been made of how, with a set of spectra in digital form, it is possible to calculate the spectra of species not directly observable. Some examples will be found in References 78 and 100–110. Knowledge of such 'terminal' spectra then allows much more chemical information concerning the system to be obtained. For example, consider an isosbestic point observed in a set of spectra recorded as a function of temperature. The next step is to establish by internal linearity checks[84] that a two-species equilibrium between, say, A and B, is present. This system may be one where the spectrum of one species free from the other cannot be recorded, perhaps because the melt solidifies before the equilibrium is displaced entirely to the left, or the melt or solute decomposes or reacts further before a high enough temperature can be obtained to force the equilibrium completely to the right and have only B in solution. If the 'terminal' spectra are calculated by established procedures,[46,106] the relative concentrations of A and B may be determined at each temperature, and hence the equilibrium constant, the enthalpy, and other thermodynamic parameters of the system become accessible.

Some pH titrations have been performed in molten salts[111] and the spectra recorded. If further studies were carried out, using digitized spectra, treating them in the way outlined by Griffiths and Potts,[112] then additional and more accurate information concerning the system would become available.

Kinetic studies in molten salts are not common. The usual spectroscopic approach is to record the change in absorbance with time at a suitable wavelength. From this information, and the absorbance when the reaction has ceased, (*viz.*, the time infinity value), the rate of the reaction may be determined. Griffiths and Pugh have shown[113] that if complete spectra are recorded at suitable time intervals,

(a) the absorbances at any wavelength can be assigned discrete times, and hence a rate constant may be determined, associated with each wavelength;

(b) neglecting values derived from wavelengths near isosbestic points or other spectral features where the absorbance changes little with time, the rate constant can be shown to be independent of wavelength, and can be presented as the mean of many determinations;

(c) the rate constants can, for a system displaying isosbestic points, be calculated by two independent methods;

(d) when the spectrum of one of the reacting species in solution cannot be measured before it starts to react (with the solvent), then its spectrum, before reaction, can be calculated; and

(e) the spectrum at the end of the reaction, the infinity spectrum, can also be calculated, and compared with the observed infinity spectrum to check for the absence of any side-reactions, and to establish whether or not the electronics of the spectrophotometer have caused drift during the course of the reaction—if they have, good reaction rates can still be obtained, using the calculated infinity spectrum.

These are examples, therefore, of some of the ways digitized spectra can and will (and should) be used in the future.

10.7. Studies Involving Thermal Gradients

Earlier sections have described how care has been taken to eliminate thermal gradients in furnaces designed for molten salt spectroscopy, but one design described a furnace with a uniform, vertical thermal gradient. The possible applications of this furnace are now outlined. The impetus for this work was a research program aimed at measuring spectroscopically the rate of dissolution of transition metal ions into a melt and across a thermal gradient, in order to understand better the hot corrosion process that occurs in a power-generation plant when superheater stainless steel tubes are attacked by molten sulfates. Existing techniques could not make continuous measurements across thermal gradients.

During the developmental stage of the technique it was soon recognized that accurate values of the diffusion and Soret coefficients of transition-metal ions in molten salts were required in order to calculate corrosion rates, and that such data were sparse. It was also realized that with the thermal gradient furnace, Soret coefficients could in principle be determined, and under the very conditions for which they were required. And diffusion coefficients could also be determined spectroscopically, using isothermal furnaces. Both approaches are essentially new techniques for determining these coefficients. They have not yet been fully tested and evaluated at this time, but they are now described.

10.7.1. Diffusion Coefficients

The diffusion coefficients of transition-metal ions in molten salts have previously been determined mainly from electrochemical measurements. This new method measures the time-dependent concentration of a transition-metal ion as it diffuses into a melt by recording the absorbance of its ligand field or charge-transfer spectrum, the system having previously been calibrated with known concentrations.

10.7.2. Soret Coefficients

Soret coefficients have only been reported once[114] for molten salt systems. Using the thermal gradient furnace, and placing in a cell an appropriate concentration of a transition-metal ion in the required molten salt solvent, the system is allowed to attain equilibrium. The concentration gradient set up can be determined from the mean of measurements of the absorbance at various points in the temperature gradient, using the narrow horizontal slot system described earlier. This gives the Soret coefficient for the transition-metal ion in the melt.

10.7.3. Dissolution Rates for Transition Metals in Molten Salts

The aggressive attack of metals by molten salts, mainly sulfates, is the fundamental feature of hot and fireside corrosion. Using the thermal gradient furnace, the rate of dissolution of a metal into a molten salt experiencing a thermal gradient (the industrial situation) can be determined. A metal coupon is placed at the bottom of a cell containing an appropriate corrosive melt and, for realistic conditions, a synthetic flue gas mixture is passed over the surface of the melt. The concentration of transition-metal ions as a function of time and distance traveled through the melt is determined next. The flux of metal ions away from the metal surface, J (moles cm^{-2} s^{-1}), is given by[115]

$$J = -Dm(\mathrm{d}N/\mathrm{d}x + SN(1 - N)\mathrm{d}T/\mathrm{d}x)$$

where D is the chemical diffusion coefficient; m, the total number of moles in a unit volume of melt; N, the mole fraction of dissolved metal; x, the distance traveled through the melt; S, the Soret coefficient; and T, the temperature (K). The metal loss rate, r, in the units conventionally used by industry, nm h^{-1}, is obtained from the flux J by

$$r = (JnA/\rho)\ 3.6 \times 10^{10}$$

where n is the number of metal atoms in the solvated complex; A, the atomic weight of the metal; and ρ, the density of the metal at the given temperature.

The necessity for good diffusion and Soret coefficient values for accurate dissolution rates is paramount. To date, using literature values, dissolution rates have been obtained which are comparable to rates determined using other techniques. However, the dissolution rate is only equal to the corrosion rate when no scale is formed on the metal surface. The extent (and nature) of scale formation may be determined from the mass of pure metal remaining at the end of the experiment combined with scanning electron microscopy, electron microprobe analysis, and X-ray diffraction studies.

11. Conclusions

The field of molten salt spectroscopy is by no means exhausted. Indeed, the current developments in the ready interfacing of spectrophotometers to mainframe computers or dedicated mini-computers, with VDU facilities now permits, with recently developed programs, further considerable and significant advances in molten salt chemistry.

ACKNOWLEDGEMENTS

Not all my co-workers have studied molten salt systems, but they have all contributed to the practice of spectroscopy at elevated temperatures and the treatment of digitized spectra in a computer. I thank them for this corpus of knowledge painstakingly (and sometimes painfully) built up over the years. Any mistakes herein from my distillation of this knowledge from their theses and oral tradition of the laboratory are mine alone. They are R. Keith Scarrow, Ran Wijayanayake, David C. Tracey, Robert (Bob) J. Armstrong, Mervyn J. Davies, Keith J. Wood, John R. Dickinson, Jim Dixon, Philip J. Potts, Richard A. Anderson, Donald C. Pugh, Hugh V. St. A. Hubbard, Keith King, and Vincent A. Creamer.

I also would like to express my sincere appreciation to the Misses Elaine Goodrich and Judith Rix for the preparation of this manuscript.

References

1. T.B. Swanson, *J. Phys. Chem.* **72,** 4701 (1968).
2. M. Kukk, *Diss. Abstr.* **25,** 1602 (1964).
3. T.R. Griffiths and K. King, in *Proceedings of the Third International Symposium on Molten Salts,* G. Mamantov, Ed., The Physical Electrochemistry Section of the Electrochemical Society Meeting, Hollywood, Florida, 5–10 October 1980, The Electrochemical Society, Inc., Pennington, N.J. (1981), Vol. 81–9, pp. 346–375.
4. T.R. Griffiths and K. King, in *Proceedings of the Third International Symposium on Molten Salts,* G. Mamantov, Ed., The Physical Electrochemistry Section of The Electrochemical Society Meeting, Hollywood, Florida, 5–10 October 1980, The Electrochemical Society, Inc., Pennington, N.J. (1981), Vol. 81–9, pp. 512–538.
5. D.M. Gruen, *Nature* **178,** 1181 (1956).
6. D.M. Gruen, *J. Inorg. Nucl. Chem.* **4,** 74 (1957).
7. B.R. Sundheim and J. Greenberg, *Rev. Sci. Instrum.* **27,** 703 (1956).
8. D.M. Gruen and R.L. McBeth, *J. Inorg. Nucl. Chem.* **9,** 290 (1959).
9. D.M. Gruen and R.L. McBeth, *J. Phys. Chem.* **66,** 57 (1962).
10. J.R. Morrey and E.E. Voiland, *Spectrochim. Acts.* **18,** 1175 (1962).
11. D.M. Gruen, *Quart. Rev., Chem. Soc.* **19,** 349 (1965).
12. C.R. Boston and G.P. Smith, *Rev. Sci. Instrum.* **36,** 206 (1965).
13. T.R. Griffiths, unpublished observation.
14. D.H. Kerridge, private communication.
15. J.P. Devlin, P.C. Li, and R.P.J. Cooney, in *Molten Salts: Characterization and Analysis,* G. Mamantov, Ed., Marcel Dekker, New York (1969), pp. 209–229.
16. R.E. Hester and K. Krishnan, *J. Chem. Phys.* **49,** 4356 (1968).
17. J.D. Mackenzie, *Rev. Sci. Instrum.* **32,** 118 (1961).
18. G.H. Wegdam, R. Bonn, and J. van der Elsken, *Chem. Phys. Lett.* **2,** 182 (1968).
19. J.K. Wilmshurst, *J. Chem. Phys.* **39,** 2545 (1963).
20. T.R. Kozlowski, *Appl. Opt.* **7,** 795 (1968).
21. J.K. Wilmshurst and S. Senderoff, *J. Chem. Phys.* **35,** 1078 (1961).
22. J.K. Wilmshurst, *J. Chem. Phys.* **39,** 1779 (1963).
23. J. Greenberg and L.J. Hallgren, *J. Chem. Phys.* **33,** 900 (1960).
24. T.S. Robinson and W.C. Price, in *Molecular Spectroscopy,* G. Sell, Ed., Institute of Petroleum, London (1955), p. 211.
25. N.J. Harrick, *Internal Reflection Spectroscopy,* John Wiley & Sons (Interscience), New York (1967).
26. J. Fahrenfort, *Spectrochim. Acta.* **17,** 698 (1961).
27. A. Bandy, J.P. Devlin, R. Burger, and B. McCoy, *Rev. Sci. Instrum.* **35,** 1206 (1964).
28. A.C. Gilby, J. Burr, Jr., W. Krueger, and B. Crawford, Jr., *J. Phys. Chem.* **70,** 1525 (1966).
29. P.A. Wilks, Jr., *Int. Lab.* **10,** No. 5, 47 (1980).
30. S.P.S. Porto and D.L. Wood, *J. Opt. Soc. Amer.* **52,** 251 (1962).

31. H. Kogelnik and S.P.S. Porto, *J. Opt. Soc. Amer.* **53,** 1446 (1963).
32. S.P.S. Porto and R.C.C. Leite, *J. Opt. Soc. Amer.* **53,** 1503 (1963).
33. J.A. Koningstein and R.G. Smith, *J. Opt. Soc. Amer.* **54,** 1061 (1964).
34. H.L. Welsh, M.F. Crawford, T.R. Thomas, and G.R. Love, *Can. J. Phys.* **30,** 577 (1952).
35. F.N.H. Robinson, *Noise in Electrical Circuits,* Oxford University Press, New York (1962), Chapter 6.
36. V.A. Maroni and E.J. Cairns, in: *Molten Salts: Characterization and Analysis,* G. Mamantov, Ed., Marcel Dekker, New York (1969), pp. 231–289.
37. C.R. Boston and G.P. Smith, *J. Sci. Instrum. (J. Phys. E.)* **2,** 543 (1969).
38. T.R. Griffiths and R.K. Scarrow, *Trans. Faraday Soc.* **65,** 1727 (1969).
39. G.P. Smith, private communication.
40. R.A. Bailey and J.A. McIntyre, *Rev. Sci. Instrum.* **36,** 968 (1965).
41. R.A. Bailey, private communication.
42. B. Cleaver, E. Rhodes, and A.R.P. Ubbelohde, *Discuss. Faraday Soc.* **32,** 22 (1961).
43. K.W. Johnson, J.R. Mackenzie, and J.N. Sandoe, *J. Chem. Soc. A,* **1968,** 2644.
44. K. King, Ph.D. Thesis, University of Leeds, 1979.
45. T.R. Griffiths, K. King, and D. Mortimer, High Temperature Technology **1,** 43 (1982).
46. T.R. Griffiths, research in progress.
47. S. Hafner and N.H. Nachtrieb, *Rev. Sci. Instrum.* **35,** 680 (1964).
48. N.H. Nachtrieb, in *Molten Salts: Characterization and Analysis,* G. Mamantov, Ed., Marcel Dekker, New York (1969), pp. 121–142.
49. S. Hafner and N.H. Nachtrieb, *J. Chem. Phys.* **42,** 631 (1965).
50. S. Hafner and N.H. Nachtrieb, *J. Chem. Phys.* **40,** 2891 (1964).
51. W.W. Warren, Jr. and W.G. Clark, *J. Sci. Instrum. (J. Phys. E.)* **1,** 1019 (1968).
52. U. El-Hanany, M. Shaham, and D. Zamir, *Phys. Rev. B* **10,** 2343 (1974).
53. U. El-Hanany, and W.W. Warren, Jr., *Phys. Rev. B* **12,** 861 (1975).
54. W.W. Warren, Jr., private communication.
55. J.P. Young and J.C. White, *Anal. Chem.* **31,** 1892 (1959).
56. D.M. Gruen, R.L. McBeth, M.S. Foster, and C.E. Crouthamel, *Abstracts of the 148th National American Chemical Society Meeting,* Chicago, Illinois, September 1964.
57. J.M. Bradshaw, Pilkington Brothers Ltd., Lathom, Ormskirk L40 5UF, England, private communication.
58. J.P. Young and J.C. White, *Anal. Chem.* **32,** 1658 (1960).
59. J.P. Young, *Anal. Chem.* **36,** 390 (1964).
60. J.P. Young, *Inorg. Chem.* **6,** 1486 (1967).
61. L.M. Toth, *U.S. A. E. C.,* 1967, ORNL-TM-2047 (Available CFSTI).
62. G.G. Cocks, J.B. Schroeder, and C.M. Schwartz, "The Spectroscopy of Fused Salts," Battelle Memorial Institute Report BM1-1185, Columbus, Ohio (Feb–April, 1957).
63. L.M. Toth, J.P. Young, and G.P. Smith, *Anal. Chem.* **41,** 683 (1969).
64. L.M. Toth, private communication.
65. J.P. Young, in *Molten Salts: Characterization and Analysis,* G. Mamantov, Ed., Marcel Dekker, New York (1969), pp. 189–208.
66. Fulmer Research Institute Ltd., Stoke Poges, Slough LS2 4QD, England.
67. V.V. Voskoboinikov, V.A. Gritsenko, V.M. Efimov, V.E. Lesnikovskaya, and F.L. Edelmann, *Phys. Stat. Sol. A* **34,** 85 (1976).
68. R.E. Hester and K. Krishnan, *J. Chem. Phys.* **46,** 3405 (1967).
69. T.R. Griffiths and K. King, *J. Chem. Soc., Faraday Trans. 1* **77,** 2763 (1981).
70. C. Solomons, J.H.R. Clarke, and J.O'M. Bockris, *J. Chem. Phys.* **49,** 445 (1968).
71. J.H.R. Clarke, C. Solomons, and K. Balasubrahmanyam, *Rev. Sci. Instrum.* **38,** 665 (1967).
72. J.H.R. Clarke and R.E. Hester, *J. Chem. Phys.* **50,** 3106 (1969).
73. C.R. Boston and G.P. Smith, *J. Phys. Chem.* **62,** 409 (1958).

74. Standard solutions of potassium chromate, and dichromate, sealed in quartz cells under nitrogen, are available from many laboratory chemical suppliers.
75. G.P. Smith and G.F. Petersen, *J. Chem. Eng. Data* **6,** 493 (1961).
76. G.J. Janz, *Molten Salts Handbook,* Academic Press, New York (1967).
77. T.R. Griffiths, *J. Chem. Eng. Data* **8,** 568 (1963).
78. T.R. Griffiths and P.J. Potts, *Inorg. Chem.* **14,** 1039 (1975).
79. W. Bues, *Z. anorg. allgem. Chem.* **279,** 104 (1955).
80. J. Greenberg and L.J. Hallgren, *Rev. Sci. Instrum.* **31,** 444 (1960).
81. H. Lux and T. Niedermaier, *Z. anorg. allgem. Chem.* **285,** 246 (1956).
82. P.J. Potts, Ph.D. Thesis, University of Leeds, 1973.
83. C.H. Liu, private communication.
84. J. Brynestad and G.P. Smith, *J. Phys. Chem.* **72,** 296 (1968).
85. F.R. Lipsett, G. Bechthold, F.D. Blair, F.V. Cairns, and D.H. O'Hara, *Appl. Opt.* **9,** 1312 (1970).
86. R.N. Jones, *Pure Appl. Chem.* **18,** 303 (1969).
87. G.J. Janz and J.R. Downey, in *Advances in Infrared and Raman Spectroscopy,* Vol. 1, R.J.H. Clark and R.E. Hester, Eds., Heyden, London (1975), p. 1.
88. W.L. Butler and D.W. Hopkins, *Photochem. Photobiol.* **12,** 439 (1970).
89. V.R. Sakalys, *Int. Lab.* **2,** 11 (1972).
90. A. Savitzky and M.J.E. Golay, *Anal. Chem.* **36,** 1627 (1964).
91. A.G. Worthing and J. Geffner, *Treatment of Experimental Data,* 5th impression, John Wiley, New York (1948).
92. J. Steinier, Y. Termonia, and J. Deltour, *Anal. Chem.* **44,** 1906 (1972).
93. J.E. Cahill, Perkin-Elmer Corporation, private communication.
94. K.E. Johnson, *Electrochim. Acta* **11,** 129 (1966).
95. S. Balt, *Recl. Trav. Chim. Pays-Bas Belg.* **86,** 1025 (1967).
96. A.P. Ginsberg and M.B. Robin, *Inorg. Chem.* **2,** 817 (1963).
97. J.R. Morrey, *Anal. Chem.* **40,** 905 (1968).
98. W.F. Maddams, *Appl. Spectros.* **34,** 245 (1980).
99. B.E. Barker and M.F. Fox, *Chem. Soc. Rev.* **9,** 143 (1980).
100. G.P. Smith and C.R. Boston, *J. Chem. Phys.* **43,** 4051 (1965).
101. G.P. Smith and C.R. Boston, *J. Chem. Phys.* **46,** 412 (1967).
102. J. Brynestad, C.R. Boston, and G.P. Smith, *J. Chem. Phys.* **47,** 3179 (1967).
103. J. Brynestad and G.P. Smith, *J. Chem. Phys.* **47,** 3190 (1967).
104. W.E. Smith, J. Brynestad, and G.P. Smith, *J. Am. Chem. Soc.* **89,** 5983 (1967).
105. T.R. Griffiths and R.H. Wijayanayake, *J. Chem. Soc., Faraday Trans. 1* **69,** 1899 (1973).
106. T.R. Griffiths and P.J. Potts, *J. Chem. Soc., Dalton Trans.* **1975**, 344.
107. T.R. Griffiths and R.A. Anderson, *J. Chem. Soc., Dalton Trans.* **1979,** 61.
108. T.R. Griffiths and R.A. Anderson, *Inorg. Chem.* **18,** 2506 (1979).
109. T.R. Griffiths and R.A. Anderson, *J. Chem. Soc., Dalton Trans.* **1980,** 205.
110. T.R. Griffiths and R.A. Anderson, *J. Chem. Soc., Dalton Trans.* **1980,** 209.
111. D.H. Kerridge, *Photoelectr. Spectrom. Group, Bull.* **17,** 508 (1967).
112. T.R. Griffiths and P.J. Potts, *Anal. Chim. Acta* **71,** 1 (1974).
113. T.R. Griffiths and D.C. Pugh, *Anal. Chim. Acta* **107,** 279 (1979).
114. V. Backlund, J. Dupuy, S. Gustafsson, and A. Ludén, *Z. Naturforsch, A* **22,** 471 (1967).
115. H.J.V. Tyrrell, *Diffusion and Heat Flow in Liquids,* Butterworths, London (1961).

5

Electrochemistry—I

F. Lantelme, D. Inman, and D. G. Lovering

This is intended to be the first part of an occasional series on electrochemical techniques in molten salts. Such a diversity of equipment, instrumental, and handling problems arises in this subject area that a division into parts seemed appropriate.

1. General Introduction

Interest in fused-salt electrochemistry has grown for several reasons, the most important being the need to free preparative electrochemistry from the limitations arising from the use of aqueous solutions. The wide potential span of ionic liquids between decomposition limits allows the electro-winning of highly electropositive metals[1] or the preparation of very electronegative elements such as fluorine;[2] their thermal stability and generally low vapor pressure are well adapted to high-temperature electrochemistry, enabling fast reaction rates to be obtained in systems which exhibit high electrical conductivity. Their ability to dissolve many inorganic compounds such as oxides, nitrides, carbides, and other salts makes them ideal electrolytic media, and extremely useful in electrometallurgy,[3] metal coating,[4] treatment of by-products[5] and energy conversion.[6]

Fused salt systems open the way to a whole new chemistry in which the relative stability of oxidation states is unfamiliar;[7] in fact, the oxidation-reduction

F. Lantelme • Laboratoire d'Electrochimie, Université Pierre et Marie Curie, 4 Place Jussieu, 75230 Paris Cedex 05, France. *D. Inman* • Metallurgy and Materials Science Department, Imperial College, London SW7 2BP, England. *D. G. Lovering* • Department of Chemistry, Louisiana State University, Baton Rouge, Louisiana 70803, U.S.A.

reactions of some compounds in molten salts at high temperatures differ from those in solvents at room temperature.[8]

On a more fundamental basis they provide an important testing ground for the theories of liquids or solutions. Indeed, when the ubiquitous nonionized solvent is removed, the Coulombic interaction becomes the predominant interparticle force and the ionic liquid can be treated as an ensemble of charged particles, whose properties can be calculated a priori by statistical mechanics.[9] Molecular dynamics calculations[10,11] lead to a better understanding of the mass transport in liquids which is one of the main features of this state: indeed, the mobility of constituent particles is a specific property of fluids and cannot be considered to derive from the mechanisms occurring in solids or gases.[12] This property is of great importance in electrochemistry for it often constitutes the limiting factor of the electrochemical reaction. This is especially apparent in the fused salt field where, in general, the electron transfer is very rapid.

Another feature of fused salt electrochemistry is the frequent use of stationary solid electrodes such as carbon, platinum, and tungsten electrodes. In general, the corresponding heterogeneous electrochemical reactions exhibit complex mechanisms. In order to obtain a better knowledge of these processes, challenging electrochemical techniques have been employed and thus the fused salt technology has played a leading part in the development of transient methods.[13,14] The recording of electrode response at a time scale of a few milliseconds proved to be valuable in examining the surface change of the electrode;[15] for example, this approach gives a good insight into the nucleation process at the very beginning of metal deposition,[16,17] or into gas bubble formation when volatile compounds are generated.[18,19] The pulse techniques at high temperatures with fused salt electrolytes constitute a unique means of studying the alloy formation at a solid electrode and of following the movement of metal atoms in the very first atomic layers[20] near the interface.

As shown in this chapter, molten salts open a wide and exciting field for the electrochemist to produce reactive elements, to study the behavior of charged species in a pure ionic medium, and, more generally, to perform physicochemical measurements in electrolytes at high temperatures. However it must be pointed out that work in this research field is by no means simple and, even more than in other fields, carelessness leads to failure or to erroneous results. The reactivity of most fused compounds is so high that traces of the usual constituents of the ambient atmosphere (especially water vapor or oxygen) must be avoided. Aggressive materials such as fluorides, fluorine, chlorine, oxides, etc. require a special environment, and furthermore the corrosion problems are emphasized by the high temperature (from 200°C up to 1500°C) of most of the experiments. Paradoxically the handling of melts is easier in large-scale cells such as industrial cells. For example, in the electrolytic preparation of aluminum[21] the continuous production maintains the CO_2 + CO atmosphere; the protection of cell walls is ensured by solid electrolyte,[22] and the ohmic heat is sufficient to maintain the temperature. These advantages have been also discussed for other production processes.[23]

These facilities do not exist on a laboratory scale since the large surface/volume ratio increases the corrosion problems, and the need for accurate measurements requires rigorous experimental conditions. After more than twenty-five years of molten salt practice, we can assert that the most important point in this research is the quality of the experimental arrangements. The control tests of materials and the trials of the different arrangements of the cell are highly time-consuming, but they are of crucial importance in meeting the experimental requirements. It is the aim of this chapter to give a survey of the different solutions which have been proposed for the most usual systems concerning electrodes, cell design, temperature control, and atmosphere purification.

2. Electrodes

2.1. Counter and Working Electrodes

2.1.1. Introduction

All the usual (e.g., aqueous solution) constraints apply to the materials selection, fabrication, and configuration for electrodes in molten salts. Thus, familiar basic principles such as the relative sizes of working and counter electrodes, symmetry, and impedance matching (particularly for high-frequency experiments), as well as the location, resistance, and junction-characteristics of, for example, reference half-cells, which pertain to less aggressive media at lower temperatures, need to be given equal consideration in melts. However, many additional criteria and limitations are also imposed. In general every aspect of the chemical, electrochemical, and thermal stability bears upon the selection of viable electrodes for molten salt operations and each melt, temperature, and kind of experiment to be performed requires individual attention to all these factors. Many detailed aspects of the materials selection, fabrication, and operation of suitable electrodes for specific applications or particular melt systems are contained within individual sections; however, the following summary of general comments and considerations may be regarded as a guide to the underlying principles.

2.1.2. Counter Electrodes

Both solid and liquid counter electrodes of many forms and materials have been utilized in molten salts. As in aqueous solutions, they must satisfy the criteria of long-term stability, relatively large mass and area, and where possible, ease and cheapness of fabrication.

2.1.3. Liquid Metal Counter Electrodes

A much wider range of liquid metals is available for use in molten salts than is normally encountered near room temperature. However, unique problems of preparation, purification, and disposition within the cell can occur for those which are not liquid at or near room temperature, i.e, most of them!

Mercury, which is obtainable in high purity and is so easily handled, has been extensively used as a counter electrode in melts investigated up to about 250°C[24] especially as a quiescent pool in the cell base. This has been especially popular for nitrate melts[25] as well as for chloroaluminates[26] in which it remains surprisingly clear and stable over long periods of time. Contact is usually made via a tungsten pin sealed in glass and inserted through the melt from above. It is less easy to bring a contact out from the base or side of the cell itself, since cylindrically symmetrical, plunge assembly configurations are normally adopted with thermostats in common use. Platinum contact wires are to be avoided with high temperature cells, because of the enhanced mutual solubility and interdiffusion of the metals.[27] Entrainment of the melt around the contact probe can be a problem if it has to be positioned through the liquid melt; sometimes it can be located before melting the salt. Pockets of melt entrapped around and beneath the metal pool, leading at best to solution inhomogeneities or, at worst, to local cells can be as much a problem as in aqueous solution. Above about 250°C, the vapor pressure of mercury becomes excessive and it tends to condense on cool parts of the assembly outside the thermostat, or to be carried as vapor on the gas purge.

Liquid lead (M. P. 327.5°C) is perhaps the second most extensively employed liquid metal counter electrode,[28] not least due to the proximity of its melting point to that of the popular LiCl–KCl (M. P. 355°C) eutectic mixture. However, some *in situ* purification is recommended, as well as preliminary clean-up procedures outside the cell.[29] The former can be achieved by a combination of gravity and differential pressure filtration (through glass frits) of liquified billets or pellets obtained from purified material previously loaded into the assembly while cold. In this way oxide skin is removed. A major problem confronting experimenters with liquid metals which melt above room temperature is that the containment cells and internal components in contact with liquid metal usually crack during cooling because of the volume changes and differential expansion stresses unless the liquid metal can be siphoned off before solidification. One expedient is to use simple, cheap replaceable containment cells, or a separate pool container such as an inverted cup, which may be raised out of the melt prior to cooling (Figure 1c). The problem is exacerbated when glass or silica is used because exchange reactions can occur with the liquid metal, de-

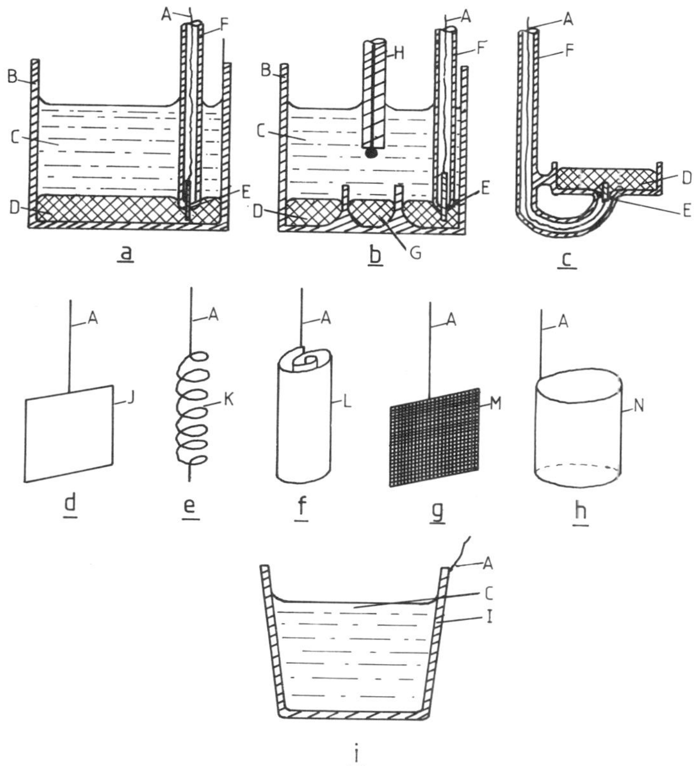

FIGURE 1. Some examples of counter electrodes: a) a simple metal pool with contact pin; b) divided metal pool; central cup isolates current-carrying pool from dropping metal, thereby reducing electrical spikes; c) inverted cup, liquid metal pool being capable of being raised, lowered, and positioned in the melt. d) plain metal foil; e) coil of wire; f) coil of metal (more suited to use inside a tubular seperator); g) metal gauze of high surface area; h) metal cylinder; i) crucible electrode. A: Electrical contact wire; B: Pyrex container; C: liquid melt; D: liquid metal; E: sealed contact electrode; F: Pyrex sheath; G: isolation compartment; H: dropping working electrode; I: metal container, J: metal foil; K: wire coil; L: coiled metal sheet; M: metal gauze; N: metal cylinder.

vitrification sets in, and the metal "tails" and sticks to the container. With semiporous containers such as graphite, the extent of mechanical damage during solidification can be severe. A less obvious difficulty, which can lead to entrapment of the melt and other spurious effects, arises in electrochemistry due to electrocapillary action. The application of a potential alters the interphasial tension between the liquid metal and the melt; if the applied signal is a rapid pulse or ramp, undesirable relaxation phenomena can occur, e.g, contact angle changes and dewetting. This is obviously a much more serious consideration for the working electrode *(vide infra)*.

Among other liquid metals, bismuth[30] (M. P. 271°C) which is adequately noble in many melts has been used. Unfortunately it has two accessible and stable valence states, making its electrochemical range of stability more limited, and it is also rather expensive. The same qualifications apply to indium (M. P. 157°C) and thallium (M. P. 303°C), though the latter is also toxicologically unacceptable. Gallium, which has the longest liquid range of any known substance (M. P. 29.8°C, B. P. 2070°C), has not met with much interest as it is costly and readily forms an oxide film upon its surface. Tin (M. P. 232°C) has a rather complicated chemistry, and zinc (M. P. 419°C) is of limited application. Silver, melting at 961°C, is adequately noble and would probably find more application in higher melting systems such as silicates were it not for its marked tendency to exchange with refractory crucible materials. Its oxide is thermodynamically unstable at the liquidus temperature. The largest and most widespread use of liquid metal pool electrodes is the aluminum cathode (M. P. 660°C) in the Hall–Héroult extraction cell.[31] Alkali metals are also formed as liquids in their molten salt extraction routes.[32]

Another serious problem encountered with molten salts is that metals having the same cation as the melt are predisposed to dissolve and give rise to significant electronic conduction. This effect has been widely studied.[33] A cautionary reminder is also appropriate concerning the high mechanical inertia which can be expected by the movement of a large mass of liquid metal. Damage and danger may not be confined to cell components where high temperatures are also involved. Anyone conversant with the operation of a Macleod manometer will testify to the possible risks.

2.1.4. *Solid Metal Counter Electrodes*

These are far more universal and much easier to use in melts than liquid counter electrodes. In their simplest form they may comprise a large plate, a spiral of wire, or a strip of gauze of some suitable noble metal, spot-welded on to a wire of the same material which is generally sheathed within an insulator. If current distribution and electrode symmetry are important a cylinder of material may be employed, surrounding the working electrode, but this reduces the visibility of the inner assembly. Provided sight of the cell components during the experiment is not too critical, the container crucible itself may become the counter electrode if it is constructed of a suitably conducting material. Naturally, the integrity of the crucible must be ensured during the experiment which means avoiding chemical reactions (e.g., with higher oxides, etc.), electrochemical potentials which dissolve the material either directly or by generated reactive intermediates, corrosive gas atmospheres, and mechanical degradation (e.g., porosity of graphite). A normal precaution would be to employ a secondary container crucible when any of the above conditions was suspected or unavoidable.

Probably the most common solid counter electrode material is platinum, although it is anodically dissolved quite readily, for example, in halides. Gold and graphite (especially vitreous carbon, now that availability is improved) are also sound choices. All three have a wide range of electrochemical stability, (i.e., both anodic and cathodic) and can be obtained at high purity in a wide range of fabrications. Since they are generally able to be used again over long periods and, in the case of platinum and gold, have a high resale value, the high initial cost may not be too serious. For highest chemical inertness at elevated working temperatures, iridium and rhodium could be adopted if the capital expenditure can be borne! Often, the best expedient is to use the best available high-density graphite which is relatively cheap and can be discarded after each experiment. This material has the added advantage of being readily machinable so that complex shapes, including threadings, are possible. Being exceptionally refractory, tungsten and molybdenum are also used in appropriate circumstances although their chemical and electrochemical stabilities are more restricted. For chloroaluminate melts a large coil of pure aluminum wire or sheet or foil is particularly appropriate and can be employed in preliminary chemical and electrochemical purification procedures (see Chapter 3, Volume 1 of this series). Recently developed reticulated (sponge) vitreous carbon is also promising and has been widely adopted as a current collector for advanced molten salt batteries.[34]

According to conventional practice, noble metals and graphite are cleaned with strong acids before use. The metals may then be flame-polished and the graphite vacuum-dried and degassed. Special techniques may be available for cleaning other materials, such as a molten nitrate dip or anodic polish in aqueous sodium hydroxide for tungsten.

Figure 1 illustrates some of the configurations possible for counter electrodes in molten salts.

2.1.5. *Working Electrodes*

Many of the comments pertinent to counter electrodes are also appropriate to working electrodes in melts, although rather more severe limitations are frequently encountered. Again, both solid and liquid devices have been used in a wide range of configurations.

2.1.6. *Liquid Metal Working Electrodes*

These need to be divided into at least two distinct categories of "dropping metal" electrodes akin to the polarographic d.m.e. and stationary electrodes, which need to be subdivided into "hanging metal drop" and other sessile drop devices, small pools, and thin films. The latter can be summarily dismissed not for their lack of utility and attraction but simply because, for some unclear reason, their potential has not been as adequately recognized as has been the case in aqueous studies.

The dropping mercury electrode (d.m.e.) itself has received considerable attention for low-melting salts, especially nitrates.[25] While all the advantages of this electrode pertaining to aqueous solutions (e.g., constantly renewed, fresh, clean surface) apply in molten salts, all the disadvantages and several more are also encountered. Probably the upper temperature limit of operation should be considered as approximately 215°C[24] although brief (<1 h, for example) temperature excursions to approximately 250°C are possible. The toxic hazard of working with mercury electrodes at elevated temperatures may be minimized by passing effluent gases from the cell through a loosely packed column of flowers of sulfur, with which it reacts rapidly and almost quantitatively.

The construction of capillaries for molten salt work requires even greater care and attention than those for use in aqueous solution.[35] Often, commercial capillaries are unsuitable because they are made of soda glass, which readily devitrifies and exchanges with melts, or because a drop-time of 3–5 seconds at 25°C becomes too rapid at elevated temperatures due to the expansion of the glass. Furthermore, common borosilicate glass may be attacked by certain ions (notably Co^{2+}) leading to tip blocking.[35] Blue-Seal glass has been found to be satisfactory.[35] At least the minimum length of capillary for uniform hydrodynamic flow should be attached directly to a wider bore of glass tubing carrying a reservoir of liquid metal, so that a significant length of it is enclosed within the cell envelope; see Figure 2.[24] Lingane[36] has clearly demonstrated that the linear flow velocity of mercury through the capillary is considerable, and that untoward consequences such as convective turbulence at the drop surface must be avoided by feeding the top of the capillary with mercury thermostatted at the same temperature as the electrolyte solution. This consideration is obviously more important in molten salts; thus the practice of terminating the capillary outside the cell, as adopted by all except one group of workers,[37] is unacceptable. Because of the raised temperature and generally short length of the capillary in most configurations, the bore of the capillary will be narrower than usual.* As might be expected, therefore, filling, cleaning, and unblocking these capillaries becomes even more of an art.

Before filling, the capillaries should be baked-out in the oven after cleaning in strong acid, so that no vapor traps exist. For the reasons explained earlier concerning the inertia of moving liquid metals, the capillary section itself is best filled by suction with the tip immersed in pure, triply-distilled clean (preferably filtered) mercury; shaking the metal into the capillary from the reservoir column can shear off the capillary joint! Naturally, capillaries need to be calibrated at the working temperature. Assuming a uniform bore, their electrical resistance

* The attachment of the fine-bore capillary length to the wider portion of tubing is best expedited by first blowing a glass bubble at one end using pressurized gas (any).

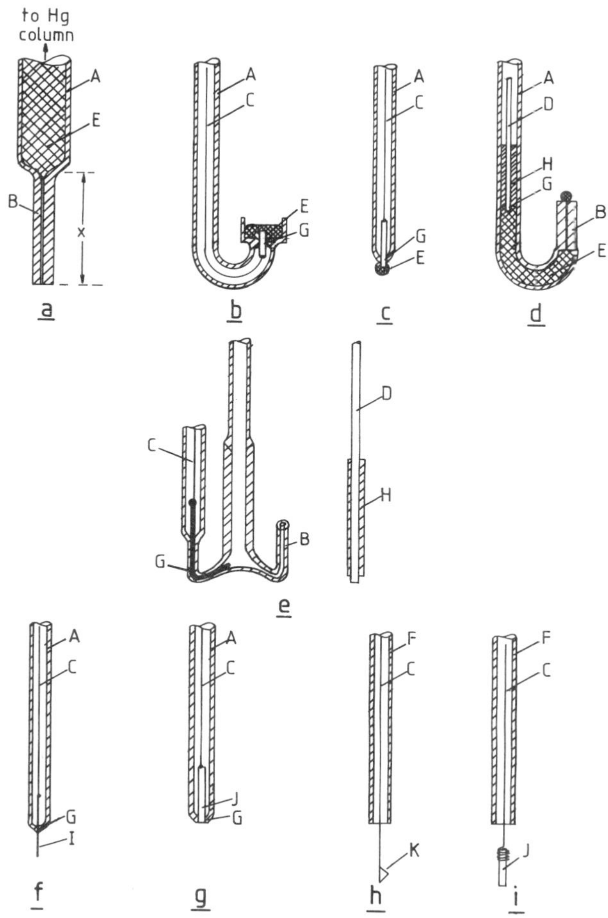

FIGURE 2. Some examples of working electrodes. a) preferred construction of capillary for dropping metal electrode in molten salts; 'x' should be at least 10 cm. for uniform hydrodynamic flow; b) upward-facing micro-cup; c) single hanging drop; d) 'upward hanging' metal drop formed by plunger attached to micrometer head; e) versatile, refillable, upward hanging, micrometer-adjustable metal drop electrode, suitable for pulse and high-frequency studies; f) wire pin; g) flush-ground; h) flag; i) rod. A: pyrex glass sheath; B: glass capillary; C: electrical contact wire; D: metal rod contact; E: liquid metal, e.g., Hg; F: refractory ceramic sheath; G: glass/metal seal and electrical contact; H: ground-glass plunger; I: thin metal wire; J: metal rod; K: metal flag.

may be measured and the bore radius determined once the capillary length is known. The rate of flow of mercury, "m", may be approximately determined (at the working temperature) as a function of total column height by collecting mercury samples issuing from the tip for known times (e.g., ten min). For more precise work, however, the capillary should be immersed in the melt and an appropriate potential applied while freely dropping samples are collected. The whole of the electrocapillary curve of drop-time *vs*. potential will need to be determined also under experimental conditions, but this can usually be achieved *in situ*. In view of the additional complexities of calibrating capillaries for molten salt work, it is worth treating them with exceptional care and, should they become blocked, it is worth spending considerable time on attempting to clear them (hot aqua regia together with evacuation while gently heating with a small flame can be reasonably effective).

In most equipment designs, the d.m.e. needs to be inserted into the cell after the system has been brought to the operating temperature. This entails momentary exposure of the melt to the atmosphere, except when the whole equipment is contained within the dry-box. An increased gas purge flow rate is therefore recommended for this procedure. If the d.m.e. is located in a flexible mount, such as an 'O'-ring set into a p.t.f.e. cell head,[38] it is possible to control the drop-time mechanically. Immediately upon removal from the cell, the d.m.e., which must remain under a significant head of mercury at all times, should be wiped clean of excess melt with a wad of glass fibers, asbestos paper, or a similar material. It should then be washed with water and/or absolute ethanol as soon as it has cooled sufficiently. Although there are varying opinions as to whether the d.m.e. tip should in general be stored beneath water, it is advantageous in molten salt work to allow it to remain in contact with water for some time so that the filament of melt that enters the bore due to the capillary response is fully leached out. For high-frequency experiments, matching the electrode assembly into the transmission line presents apparently unsolved problems; the usual constraint on design of cylindrical symmetry, *(vide supra)*, often implies that the current and signal inputs must be made at the same point or very close. This limitation can be removed with stationary electrode designs.

Apart from mercury, dropping gallium,[39] bismuth,[30] and silver[40] electrodes have been described. In the latter case, a miracle of technology in itself, the lifetime of the silica capillaries was limited by the diffusion of silver into their structure. The dropping bismuth electrode was refined for reproducible and reliable operation by one author[30] but few others have taken up the challenge.

Much greater flexibility in design is realizable for stationary liquid metal working electrodes. The simplest concepts of all in this direction are the upward-facing micro-cup (Fig. 2b) and the single hanging drop located on a contact pin, usually platinum (Fig. 2c). In the former case, it is difficult to renew the small pool and in the latter case this may or may not be accomplished according to the construction, the metals used, and the overall cell versatility. More attention might usefully be directed to liquid metal thin-film electrodes.

The next stage in sophistication is a hanging metal drop electrode for which the drop may be reproducibly renewed. The only commercial design, suitable for mercury (and probably gallium), is the Radiometer P958b. The essential parts of this device are constructed of glass, stainless steel, p.t.f.e., and epoxy resin—the cosmetic plastic components may be discarded. Thereafter, the electrode can give good service over long intervals up to 250°C, with only periodic replacement of the capillaries. Various alternative designs of plunger or micrometer screw hanging mercury drop electrodes have generally proved unsatisfactory except in the hands of their originators.[41] Several elementary designs of "upward hanging drop" have proved applicable to a range of liquid metals and melts over a wide range of temperatures: a number of drops may be expelled in turn by the operation of a plunger, normally connected to a micrometer head.[25] The limitations of these laboratory-constructed devices include their restricted reservoir capacity, and the irreproducibility of their drop areas; filling and emptying may also present problems.

An unusually versatile "downward hanging" metal drop electrode, based on an earlier "upward hanging" version (see Fig. 2e), was designed by Graves and Inman, improved by Lovering, and refined by White over a period of many years.[42] Although originally constructed of borosilicate glass (with internal precision-ground sections), tungsten, and a remote brass micrometer head, it could equally well be made with silica and platinum or even iridium contacts. Using this design, the capillary and small reservoir could be repeatedly filled (and subsequently emptied) from the large liquid metal pool at the base of the cell by lowering the pressure in the main assembly, plunging the tip beneath the pool, and returning the cell to operating pressure. Although some skill and dexterity of operation is called for, the success of this device has been demonstrated with a range of metals and melts. It is particularly suited to high-frequency and large-current pulse experiments, being well-matched into the transmission line and exhibiting only a small transient overshoot without ringing.

The calibration of drop areas is essential with all the above designs. For liquid metal drops, the effective electrochemical area may be taken as the geometrical area, and hence is calculable from, for example, microscopic or photographic measurements. Since these will be performed at room temperature, appropriate allowance must be made for the expansion of the orifice at operating temperature. If drops hang downwards, they may be mechanically dislodged, collected, and characterized by weighing when cool. Providing sufficient data is available, operational areas may be checked electrochemically through the measurement of the diffusion limited current recorded for a tracer ion. It may be necessary to allow for any expansion or contraction of the drop at the working potential because of electrocapillarity.

Attention also needs to be paid to the correction of standard formulae in subsequent calculations to ensure that all the consequences of working at the elevated temperature are accounted for.[43] This applies especially to those equations derived on the basis of the approximate equality of diffusion rates in the electrolyte phase and the liquid-metal phase. Diffusion in liquid metals is generally much faster at higher temperatures, while ion interdiffusion in melts may be slower than in aqueous solution at room temperature but exponentially related to temperature. The ratio of the two diffusion coefficients which frequently occurs in derivations may not, therefore, be approximated to unity.

Whilst the mutual solubilities of metals are enhanced at raised temperatures (notably when one is liquid), the solubility limit may be exceeded in electrodeposition experiments for some metal combinations or at high current densities at the electrode surface. This may lead to precipitation and spurious responses. Alternatively, in strongly oxidizing melts deposited metal may undergo rapid oxidation at the surface, leading to irreproducible behaviour. With a stationary liquid metal electrode such effects may be less apparent (but none-the-less important) than with a dropping metal electrode.[44]

2.1.7. Solid Metal Working Electrodes

This constitutes the largest group of working electrodes selected for molten salt studies, primarily due to overriding practical limitations of all the alternatives in most systems.

In some very early studies[45] attempts were made to operate gas-flushed, noble metal electrodes in which the surface was periodically exposed to the melt. It was hoped that the electrical response would simulate that of a dropping metal electrode which was indeed the case in some studies. However, the ultimate limitation of all solid electrodes, contamination, eventually restricts the lifetime of their reproducibility or predictability. In addition, their complexity of design and operation, as well as the lack of any theoretical basis, precludes their serious consideration.

Generally, the construction of solid electrodes has been rather elementary (Figure 2). At lower temperatures ($<450°C$), at which borosilicate glass forms a suitable insulator, a noble or refractory metal such as platinum, gold, tungsten, or molybdenum may be sealed into one end of a glass tube. This may be used as it is (Figure 2f), or the end may be ground and polished flush. In either case, the area approximates to the geometric area, provided the whole of the end of the electrode is immersed beneath the melt. The electrochemical area may also be determined with a suitable tracer ion in order to assess the "roughness" factor. The sealing of platinum and gold into borosilicate glass usually leads to difficulties because of the wide separation between their expansion coefficients. Invariably, melt soon creeps between the glass and the metal, leading to unreliable data (a sensible maxim in molten salt electrochemistry is to be suspicious of any results which suggest a non-diffusion-controlled phenomenon, until evidence can be mustered to confirm it). The use of Supremax glass and silica sheaths does not alleviate this problem but introduces additional problems of fabrication due to their higher softening points. In spite of the indeterminacy of their (electronically-conducting) surface oxide compositions, tungsten and molybdenum, which can be successfully sealed into a wide range of glasses, have found wide application as microelectrodes for melts. For nonoxidizing melts (and even nitrate melts at very low temperatures, e.g., $<150°C$), vitreous carbon working electrodes are invaluable, and they may be rim-sealed into the lower-melting glasses.

At the higher temperatures (say, $\gtrsim 650°C$) and for the most aggressive melts (e.g., fluorides, oxides, etc.), a small flag (Figure 2h) or rod (Figure 2i) attached to a wire, freely located (at the lower end) within an insulating refractory ceramic sheath, may be used. Only the flag or the tip of the rod is immersed beneath the melt in these cases. The area may be determined *in situ,* electrochemically, or subsequently and less reliably by examination of the wetted portion during the experiment.

With flag electrodes, rather fine wires are attached so that the immersed area approximates to that of the flag itself. Of course, electrocapillary effects may again cause variation of the wetted area during an experiment, especially with rods and wires.

The adoption of signal-reversal transient techniques such as cyclic voltammetry[46] and current-reversal chronopotentiometry[47] have considerably enhanced the lifetimes and improved the reliability of solid electrodes in melts. According to these methods electrodeposited metal films, present in a very active state, may be reoxidized immediately before chemical reaction with the melt occurs; similarly, rapid re-reduction of gas films formed anodically may hinder their chemical reaction with the melt or electrode substrate. Nevertheless, after considerable use, the cumulative effect of these procedures may produce a "harrowed" electrode surface, which consequently exhibits a large "roughness" factor.

Another cautionary note regarding product interaction with electrodes is appropriate. Johnson[48] has repeatedly warned about the cathodic corrosion of metal electrodes in nitrate melts and further studies[49] have indicated that the reactive intermediate generated during the electroreduction of the nitrate ion is capable of chemically dissolving all metals and alloys, regardless of how noble, passive, or otherwise resistant they might normally appear.

Not many workers have attempted to operate rotating electrodes in molten salts.[50] The problems of sealing the driven section from the environment are severe, but apparently not insoluble. The additional problems of insulation at elevated temperatures have deterred all except one group[51] from constructing a ring-disc electrode, but those authors possessed facilities for conducting their experiments wholly within a dry-box—a level of sophistication not enjoyed by many.

An interesting consequence of studying the electrodeposition of metals from molten salts results from the varying state and degree of interaction of the deposit with the electrode substrate.[52] As indicated earlier, many metals may be liquid at the experimental temperature and, that being so, a wider range of alloying, compound-formation, and interdiffusion reactions than usual are feasible. For this reason, the selection of appropriate electroanalytical equations describing the observed behavior needs special consideration. For example, the electrodeposition characteristics of zinc (M. P. 419°C) from a LiCl–KCl eutectic melt at 400°C will not only be different from those at 450°C (both common working temperatures with this system), but will also change when the substrate is, for example, tungsten (an insoluble deposit) rather than platinum (a soluble deposit).

For specialist applications, other working electrodes have been devised. Some examples are the development of optically "transparent," porous glassy carbon, vapor-deposited thin film, and metal grid electrodes for spectroelectrochemical studies, described in greater detail in the Chapter by Norvell and Mamantov (Volume I). Semiconductor electrodes have only recently been investigated in melts, but offer some interesting prospects.[53] As working electrodes, per se, fast ion-conducting materials, electronically-conducting interstitial alloys, and metal oxides have yet to be exploited.

2.2. *Reference Electrodes*

2.2.1. *Introduction*

In spite of formidable experimental difficulties many reversible electrodes have been designed for use as thermodynamic activity *or* concentration probes in melts at high temperatures in excess of about 100°C. Several very useful general surveys were published some years ago[54] at a time when many new reference electrodes were being introduced. Therefore only the theoretical and practical principles governing the application of these electrodes in melts and the details of a few of the more widely-used types will be given in this chapter.

It is often difficult to design suitable electrodes for high-temperature applications because of the problems which arise from, for example, side reactions such as the corrosion of metal electrodes and the glass or ceramics used as containers or diaphragms, very rapid electrode reactions involving impurities, and the increased solubilities of the salts normally used in electrodes of the second kind. Unsuspected and undetermined liquid junction, thermoelectric and thermal diffusion potentials produce uncertainties in the measurements and render them less precise and reproducible than those in aqueous solutions. In addition, temperature gradients lead to the frequently observed redistribution of the metallic components of the systems. Water, as an impurity, can have particularly deleterious effects. For example, chloride melts containing lithium ions are particularly easily hydrolyzed, with the formation of oxide ions and HCl.

There is no fundamental difference between the two half-cells or electrodes which comprise a cell whose emf is measured. This cell symmetry is particularly marked in molten salt systems. However, it is usual to designate one of them as the *reference half-cell or electrode* if it is of such general applicability that it can be used as the reference point for an emf series. In many cells it is necessary to separate the two half-cells and in these cases the consequent diaphragm is conveniently regarded as part of the reference half-cell. If the measured emfs are to be used to calculate thermodynamic quantities then liquid junctions must be absent or, if present, the consequent liquid junction potentials must be known or eliminated (see later). The various inert substrates such as platinum and mercury which are used in the study of electrode processes and which are generally known as indicator electrodes are considered in another part of the chapter.

Emf series are more extensive in halide melts than in oxy-anionic melts. In the latter melts, the oxy-anion may be cathodically active at potentials which are much less cathodic than those at which the solvent cations (generally alkali metal ions) are reduced to metals. The available ranges of solvent electroinactivity are also less than expected in halide melts because alkali metals for example are soluble in their own salts, but this is a much less marked effect than that in the oxy-anionic melts.

Some emf series which illustrate this point are given in Table I. In addition, metal ions and in particular noble metal ions are frequently acidic enough (in the Lux–Flood sense) to abstract oxide ions from the oxy-anions (e.g., $2Al^{3+} + 3SO_4^{2-} \rightarrow Al_2O_3 + 3SO_3$.[55] Thus, although the *metal* may co-exist with the solvent melts, the metal-metal ion couple may not exhibit a stable electrode potential. The noble metals that behave in this way may of course be used to monitor the redox potentials of melts.[56] In this mode they act as inert electron sinks for the various oxidation–reduction reactions which interact to determine the redox level of the melt.

The standard hydrogen electrode which is employed as the primary reference

TABLE I. Standard Reduction Potentials (Volts) on the Ag°/Ag^+ (Mole Fraction) Scale

Solvent Melt	LiCl–KCl	NaCl–KCl	
Temperature	450°	700°	Li_2SO_4–Na_2SO_4–K_2SO_4
Au/Au^+	+0.948		
Cl_2/Cl^-	+0.853	+0.845	
Cu^+/Cu^{2+}	+0.682	+0.596	−0.099(575°)
Cr^{2+}/Cr^{3+}	+0.006	+0.246	
Ag/Ag^+	0.000	0.000	0.000
Ni/Ni^{2+}	−0.158	−0.140 (at 670°)	
Cu/Cu^+	−0.214	−0.256	−0.220(580°)
Co/Co^{2+}	−0.354	−0.319	−0.764(550°)
Pb/Pb^{2+}	−0.464	−0.390	
Fe/Fe^{2+}	−0.534	−0.520	
Cr/Cr^{2+}	−0.788	−0.758	
Zn/Zn^{2+}	−0.929	−0.860	
Mn/Mn^{2+}	−1.212	−1.206	
Li/Li^+	−2.773		

electrode in water and other protonic solvents is not generally applicable in ionic melts although some melts do contain the hydrogen ion or the hydride ion as a main component (for example, the molten bisulphates[57] and hydrides.[58]) Also, dilute solutions of acids such as HCl can be prepared in molten alkali metal chlorides and one investigation, for example, has concerned the hydrogen electrode in the HCl–LiCl–KCl melt.[59] The most widely applicable electrode of this class in ionic melts is the simple oxygen electrode (O_2/O^{2-}) and its modifications (e.g., $CO_2/O_2/CO_3^{2-}$) for use in carbonate melts, although difficulties often arise in practice from the corrosion and oxidation of the supposedly inert metal substrates on which these electrodes are established. In molten halides, the obvious primary reference is the corresponding halogen/halide electrode, for example the chlorine electrode in molten chlorides.

These electrodes, along with some of the secondary reference electrodes which are often more convenient in practice, are discussed in the following sections.

2.2.2. Quasi-Reversible Electrodes

The simplest type of reversible electrode is the so-called quasi-reference electrode.[60] When a piece (area 1–2 cm^2) of a noble metal such as platinum is dipped into a melt, it generally assumes a steady electrode potential which remains approximately constant as long as the gross composition of the melt is not changed. The platinum electrode, in the absence of platinum ions, will of course monitor the redox potential of the melt, which, if unchanged, causes the potential of the platinum electrode to remain constant. An alternative explanation of the mechanism of operation of the electrode is that the small quantity of platinum ions formed in the melt by corrosion may strongly adsorb on the electrode surface and give rise to a well-defined adsorption potential.[61] This type of electrode has found important applications as the large, virtually unpolarized, counter (reference) electrode for voltammetric and chronopotentiometric studies in melts.[62]

Nickel and platinum wires have been used as quasireference electrodes in molten fluorides but as in other melts their potentials are not well-defined (but stable to ± 10 mV for a period of months according to Mamantov.[63]

2.2.3. Gas-Gas Ion Electrodes

In this class, the chlorine electrode has been widely applied[64] because molten chlorides have been very popular experimental media. However, it is not easily set up (the construction of the electrode is rather complex) and it cannot be used in those melts where the solute is oxidizable. For example, it cannot be used for the study of solutions of uranium trichloride (UCl_3) or titanium trichloride

($TiCl_3$) in molten alkali metal chlorides unless separated from the bulk of the melt because chlorine gas oxidizes them to UCl_4 and $TiCl_4$ respectively. The electrode generally takes the form of a graphite rod or tube saturated with chlorine gas, either from the cylinder or prepared *in situ* by anodic evolution. The latter procedure can only be employed in pure chloride melts. The graphite must be carefully pretreated to remove impurities. With suitable precautions it has been possible to obtain results with the chlorine electrode which agree with the values calculated on the basis of thermodynamic quantities obtained from nonelectrochemical studies.

The bromine/bromide and iodine/iodide electrodes are set up in a similar manner.[65] Murgulescu and Popescu[66] in particular gave useful experimental details for a reversible bromine electrode.

Oxygen reference electrodes[67] are necessary and convenient for the study of the thermodynamic and electrochemical properties of oxyanionic melts. In principle the electrode consists of an inert, noble-metal substrate (carbon is not inert in the presence of oxygen) which functions as the electron sink, bathed by oxygen or air and dipping into a melt containing oxide ions. The electrode has been used in many forms, but electrodes employing platinum substrates have been the most widely used, e.g., in molten sodium hydroxide and in sulfates, carbonates, silicates, borates, phosphates, and cryolite-alumina melts. Oxygen gas presumably gives rise to adsorbed oxygen atoms on the platinum or platinum oxide substrates. These atoms are then in equilibrium with the oxide ions in the melt. The latter arise from Lux-Flood acid-base equilibria of the type:

$$SiO_3^{2-} \rightleftharpoons SiO_2 + O^{2-}$$
$$PO_4^{3-} \rightleftharpoons PO_3^- + O^{2-}$$
$$P_2O_7^{4-} \rightleftharpoons 2PO_3^- + O^{2-}$$
$$B_4O_7^{2-} \rightleftharpoons 2B_2O_3 + O^{2-}$$
$$CO_3^{2-} \rightleftharpoons CO_2 + O^{2-}$$
$$SO_4^{2-} \rightleftharpoons SO_3 + O^{2-}$$

The oxide activities in the melts are determined by the appropriate equilibrium constants. Although in principle all these melts should be acid-buffered to prevent thermal decomposition (e.g., by CO_2, a Lux-Flood acid, in the case of carbonate melts), some of them, such as the sulphates, are very thermally stable and are often used under oxygen *or* inert gas atmospheres.

The oxygen electrode generally gives rise to a stable potential but is sometimes somewhat sluggish in operation. The rate of flow of oxygen and the position of the wire substrate may also affect the electrode potential. The potential of the oxygen electrode should be the following (Nernstian) function of the oxygen pressure and oxide ion activity:

$$e = e_0 + \frac{RT}{2F} \ln \left\{ \frac{p_{O_2}^{\frac{1}{2}}}{a_{O^{2-}}} \right\}$$

which is deduced on the basis of:

$$O_2 + 4e \rightleftharpoons 2O^{2-}$$

as the overall electrode reaction. Experimental behavior in accord with the equation above also evidences the reversibility of the electrode. In basic melts slopes of *RT*/F rather than *RT*/2F ions have been observed for the plots of electrode potential *vs.* the logarithm of oxide ion activity (generally given approximately by the concentration, at least in dilute solutions).[68] Several attempts have been made to explain these ambiguous slopes. One such explanation involved[69] the formation of peroxide ions through the reaction

$$^1/_2\, O_2 + O^{2-} \rightarrow O_2^{2-}$$

However, it is more likely that the peroxide-forming reaction takes place heterogeneously by means of two coupled electrode reactions at the metal electrode surface. Thus we have

$$\left.\begin{array}{l} ^1/_2\, O_2 + e \rightarrow O^- \\ \\ O^{2-} - e \rightarrow O^- \end{array}\right\} O_2^{2-}$$

with the electrode exhibiting a mixed (steady-state) potential. However, as the homogeneous redox couple O^{2-}/O^- is likely to have a much higher exchange current density than that involving oxygen gas, it will essentially control the electrode potential which will be given by the equation:

$$e = e_0 + \frac{RT}{2F} \ln \left\{ \frac{a_{O_2^{2-}}}{a_{O^{2-}}^2} \right\} = e_0 + \frac{RT}{2F} \ln a_{O_2^{2-}} - \frac{RT}{F} \ln a_{O^{2-}}$$

which is deduced on the basis of the reaction:

$$2O^{2-} - 2e \rightleftharpoons O_2^{2-}$$

as the potential-determining electrode reaction. In this way it is possible to account for the *RT*/F slope. The dependence of electrode potential on oxide ion activity in buffered oxyanionic melts is not anomalous because the oxide ion concentrations in this case are too low for peroxide ions to be formed. As briefly mentioned above, in oxyanionic melts such as carbonates, in which the oxyanion easily dissociates, the acid component (CO_2) must be mixed with the oxygen in setting up the reversible electrode so as not to disturb the acid–base equilibrium:

$$CO_3^{2-} \rightleftharpoons CO_2 + O^{2-}$$

Reference electrodes for molten carbonates have recently been discussed extensively by Selman and Maru,[70] so only the main points will be mentioned here.

For carbonate melts, the oxygen electrode reaction (necessarily in the presence of CO_2) becomes

$$CO_3^{2-} \rightleftharpoons CO_2 + {}^1/_2O_2 + 2e$$

and the most widely used primary, gas–gas ion, reference electrode is based on this reaction.

An alternative reference electrode is based on the reaction

$$CO_3^{2-} + CO \rightleftharpoons 2CO_2 + 2e$$

but the disadvantages of this electrode are obvious.

The most widely used reference electrode (utilizing platinum or gold substrates) was thoroughly tested by Janz and Saegusa.[67] It was shown to be stable and reproducible (within 5 mV) after it was aged for about 20 h. One disturbing feature was that the electrode potential depended on the substrate employed. As oxide layers seem to play a role in determining the potentials on platinum (and platinum-related metals), gold has been adopted as the generally applied substrate material for the oxygen reference electrode at 800–1100 K. Graphite has also been employed but the behavior is then complex because of reactions between oxygen, oxide ions, and the graphite substrate.[71]

2.2.4. Metal-Metal Ion Electrodes of the First Kind

Many electrodes of the first kind involving metals in equilibrium with melts containing the corresponding metallic ions have been employed and in principle their application is quite straightforward. These electrodes are easily prepared and, because of the high exchange current densities of metal–metal ion couples in melts, rapidly come to equilibrium, are not easily perturbed, and are highly reproducible. However, when used as reference half-cells, the metal–metal ion electrodes must be isolated, but not insulated, from the bulks of the melts, and major problems often attend this separation. In general, apart from the practical difficulties, this set-up necessarily introduces unknown liquid junction potentials, although these may be minimized in certain circumstances.

The example of the lead electrode may be cited to illustrate this class of electrode.[72] It is simple to construct and is easily reversible. The low melting point of lead and its relatively low vapor pressure are additional advantages. Also the surface of liquid lead is very reproducible. The solubility of lead in molten lead chloride is also low.[73] The problems of preventing intermixing and minimizing liquid junction potentials will be discussed later.

The sodium electrode was used, for example, by Delimarskii and his co-workers.[74] Emf series established on the basis of this electrode in different sodium ion *or* sodium ion-containing salt melts could perhaps be meaningfully compared if the sodium–sodium ion potential were to a first approximation solvent-independent (*cf.* use of rubidium electrode in nonaqueous, organic solvents[75]). Although complex formation between the sodium cation and the anions of the solvent is probably negligible, because of the relatively low polarizability and polarizing power of the former species, cation–anionic electrostatic forces will depend on the anion and will probably invalidate this approximation. However, pure sodium is appreciably soluble in melts containing sodium ions and non-reducible anions[76] and reacts vigorously with melts containing reducible anions. Thus it is necessary to dilute the sodium in an inert, relatively noble, metal solvent and/or interpose a solid electrolyte such as glass or β-alumina, which conducts by the transport of sodium ions, between the metal and the melt.[97] Thus the electrode may be formulated as:

$$\text{Na,Sn(or Hg)} \;\Big|\; \underset{\text{ion conductor}}{\text{solid Na}^+} \;\Big|\; \text{Na}^+$$

Sodium amalgams (Na–Hg) have been used up to 300°C.[78] Soft glasses are used at the lower temperature, Pyrex glass can be used up to 500°C, but quartz glass is recommended for higher temperature applications.[79] A review on molten salt–glass interactions also deals with applications in cells.[80]

Pt/Pt^{2+} electrodes may be used in melts. Laitinen and his co-workers[81] in particular adopted this electrode as a secondary reference electrode in the molten LiCl–KCl solvent. The Pt^{2+} concentration is usually ~0.01 *M*. Other workers have adopted the Ag/Ag^+ electrode for similar purposes.[82] An obvious advantage which both these electrodes possess is that the ionic component may be prepared in situ by anodic dissolution in most melts. Although platinum is the more noble metal and thus may be employed in contact with a greater range of ions without displacing them from the melt, it suffers from the disadvantage that oxides easily form on the platinum surface and that the platinum halides are not as thermally stable as the silver halides. Silver oxide also decomposes at a fairly low temperature.

Mercury has been employed in the form of pools as a quasireference electrode,[83] and in mercury–calomel[84] and mercury–mercurous sulfate[41] electrodes in various low-melting nitrate mixtures.

2.2.5. *Electrodes of the Second Kind*

These have only found a limited use in melts, e.g., in nitrates at very low temperatures,[86] mainly because the solubilities of most inorganic compounds are high in melts. For example, the silver–silver chloride electrode may be employed to determine the concentrations of chloride ions in low-melting nitrate eutectics based on lithium nitrate,[87] but it is not so useful in the sodium nitrate–potassium nitrate melt (at, say, 250°C) because of the greatly increased solubility product of silver chloride at the higher temperature. Metal–metal oxide electrodes may have a wider applicability (they are especially useful in those cases where the oxygen electrode cannot be used because the exchange current density is too low or because oxygen reacts with the oxide ions under investigation to form peroxide ions). Extensive investigations of metal-metal oxide electrodes were carried out by Laitinen and Bhatia in the LiCl–KCl melt at 450°C.[88]

2.2.6. *Electrodes of the Third Kind*

Metal oxides have also been employed in electrodes of the third kind in melts. This type of electrode is usually very sluggish in operation in aqueous solutions but the higher intrinsic exchange current densities in melts make them less so in these media even though the concentrations of potential-determining ions in equilibrium with the rather insoluble oxides are very low. However, they still respond somewhat slowly to changes of metal-ion concentration, presumably because the rates of precipitation and dissolution of the component oxides are still slow at high temperatures. The Pd/PdO/CdO electrode has been employed in studies of complex-ion formation to monitor Cd^{2+} concentrations in molten nitrates.[89]

2.2.7. *Specific Problems Concerning Reference Half-Cells*

As indicated in the introduction to this chapter, there are three main problems involved in devising reference half-cells.

a) There should be no chemical or physical interaction between the metal electrode and the melt. In addition, the exchange-current densities of the metal–metal ion couples should be sufficiently high to ensure adequate reversibility. Low exchange-current densities are generally not a problem in melts, at least with metal–metal ion couples, and chemical and physical interactions are minimized by employing relatively noble metals. Oxidizing impurities should certainly be avoided because in these circumstances some of the supposedly inert metals may corrode at an appreciable rate. Temperature gradients should also be avoided.

b) If the reference electrode is to be used for determining thermodynamic quantities or for establishing an emf series, the liquid junction potential between the melt in the reference half-cell compartment and the bulk melt must be eliminated or be known and constant. In other applications, it may only be necessary for the liquid junction potential to be constant during any sequence of measurements.

c) The melts in the reference half-cell and the bulk should not intermix, but there should be adequate ion conduction between the compartments. That is, the reference half-cell must be isolated but not insulated from the main part of the system.

These requirements are best satisfied by the type of reference half-cell shown below:

M	MX	(dilute solution) (in appropriate) (solvent)	| diaphragm

The choice of an M/M^{n+} couple (M = Ag or Pt) insures that the interaction between metal and melt is minimal and that the exchange-current density is high. The diaphragm may be continuous or porous (see below). The former type is preferred because of its electrochemical and physical stability and because it rigorously prevents intermixing. The use of a dilute solution in the reference electrode compartment insures that the liquid junction potential is close to zero, on the assumption that the solution–solid membrane potentials on the two sides of the diaphragm are equal and opposite.

In the case of a porous diaphragm if the solutes in the half-cells on both sides of the diaphragm are present as dilute solutions in the same solvent (the ions of which conduct practically all the current through the cell), then the liquid junction potential is virtually zero as can be seen by considering the equation:

$$E_{\mathrm{LJP}} = \frac{-RT}{F} \int_{A}^{B} \sum \frac{t_i}{z_i} \cdot \mathrm{d}\ln a_i \rightarrow 0$$

(the symbols have their usual meanings) because the concentrations of the ions which conduct the current are virtually the same in both half-cells. Thus on a quasithermodynamic basis this reference half-cell can be used safely for the precise intercomparison of electrode systems in the same solvent melt and for thermodynamic studies, although it can only be applied when the solutions are dilute in the latter case. Ultimately, and especially when the solute concentrations are relatively high, irreversible processes of interdiffusion lead to time-dependent potentials.

Thus a continuous diaphragm is preferred. For example, glass-contained electrodes have frequently been employed in low-temperature melts such as nitrates and chlorides.[90] If a glass bulb is used to contain the reference half-cell, then the boundaries between the melt in the reference half-cell and the inner surface of the glass and the bulk melt and the outer surface of the glass can be regarded as sharp liquid junctions. Therefore, if sodium ions are the only mobile species in the glass and the melts contain sodium ions as a major component, the overall junction potential of the glass diaphragm, obtained by integrating equations of the type:

$$E_j = \frac{-RT}{F} \int_A^B \frac{t_i}{z_i} \cdot d\ln a_i$$

for each boundary (where the symbols have their usual meanings), should be zero and the measured emf should depend only on the difference between the solute metal ion concentrations in the two halves of the cell for dilute solutions (assuming that the appropriate metal electrodes are employed in the two halves of the cell). Small asymmetry potentials are sometimes observed,[91] even when the compositions of the melts on both sides of the glass diaphragm are the same. In many measurements, it is only necessary that these asymmetry potentials should be constant and this is generally so. They are probably due to nonuniform glass compositions and can be removed by careful annealing of the glass.

The *diaphragm materials* are chosen on the basis of the experimental temperatures and the corrosive natures of the melts. Glass, silica, and silica–alumina minerals such as mullite may be employed in acidic melts over specific ranges of temperature up to about 1400°C.[92] However, these materials are corroded in basic melts (i.e., those with low pO^{2-} values) and boron nitride, magnesia, and zirconia are more likely to succeed in these melts.[93] Alumina, on account of its refractory nature, is also fairly unreactive.[94] Graphite (for use as porous diaphragms only) is useful in molten fluorides,[95] melts of which pose particularly difficult problems (see below).

Although *porous* diaphragms are subject to the limitations discussed above, they have found many applications. The following types have been employed: a) fritted discs made of glass, ceramics, or graphite;[96] b) asbestos plugs placed in narrow tubes joining the two halves of the cell;[97] c) asbestos fibers sealed into glass tubes;[98] and d) simple capillaries.[99] Of these, the asbestos fiber type has proved to be the most successful but some workers have experienced difficulties.[100]

As discussed above, glass has been used in *continuous* diaphragms. The conductance of the glass is sufficiently high at temperatures $\gtrsim$200°C for low-impedance potentiometers to be employed for the emf measurements. For example, a Pyrex glass bulb approximately 1.25 cm in diameter and 0.1 mm thick had a resistance of 16 kΩ at 250°C. This bulb was fairly robust. If more robust electrodes are necessary, then high-impedance potentiometers must be employed. These glass electrodes are of course limited to low ($\lesssim$600°C) temperatures, although modifications employing silica or high-silica-content glasses have been described for higher temperature applications.[101]

These silica-containing materials cannot, of course, be employed in basic melts. Alkali metal ion-conductive ceramics are useful in these latter systems although there is some evidence for the transport of Ag^+ ions through some of them.[102] Many French workers have employed Pythagoras porcelain[103] but Labrie and Lamb[104] prepared their own porcelain. Mullite exhibits sufficient alkali metal ion conductivity to be useful in this respect (as tubes) although high-impedance potentiometers must be used for the emf measurements.[105] β-alumina has also proved to be useful (see above). Fluoride melts are particularly corrosive; some useful reference electrodes for these systems are fully described in the monograph by Alabyshev, Lantratov, and Morachevskii.[54]

The *solid electrolytes* which conduct electricity by the migration of oxide ions (e.g., ZrO_2–CaO mixtures) are also being increasingly used for emf measurements in molten salts.[106] They may, of course, be employed as oxide ion-concentration-responsive electrodes to measure pO^{2-} (the use here is similar to that of glass to monitor pH in aqueous solutions) or as diaphragms in systems where pO^{2-} remains approximately constant in the melts on both sides of the diaphragm while other factors change, and in the cases where the use of the primary oxygen electrode (say, on platinum substrates) is often not possible because of conjugate corrosion and oxidation processes and peroxide or superoxide formation.

The development of reference half-cells for use in molten fluorides deserves a special mention in this context as it has posed particularly difficult materials problems. The Ni(II)/Ni electrode isolated but not insulated from the main system in a separate compartment has been used by several authors.[107,108] The major drawbacks of such electrodes lie in the materials used for the diaphragms. For example, boron nitride, a popular material, is apparently attacked by molten fluorides. Although in many cases it is likely that this is merely the leaching out of the boric oxide used as the binder, cryolitic (aluminum ion-containing) melts are known to have a corrosive action on boron nitride itself.[109] Evidence for the formation of aluminum nitride as well as sodium boron nitride, $NaBN_{13}$, which is presumably an intercalation compound, has been found but the exact mechanism of the corrosive action is unknown.

Presumably these boron nitride membranes are sufficiently porous (after leaching) and wetted for there to be a 'molten salt bridge' between the compartments through the diaphragm. More overt attempts have been made to establish well-defined ionic junctions between the compartments. Porous silicon nitride sheaths (20 mesh) have been tried but unfortunately the material is rapidly dissolved, at least by cryolitic melts and pure sodium fluoride.[110]

Calcium-stabilized zirconia diaphragms (ionic conduction through oxide (O^{2-})ions) were satisfactory for up to 4 h in sodium fluoride, but did not last for more than 30 min in cryolitic (AlF_3-containing) melts even when these were saturated with oxides.[110] These diaphragms would of course only be useful for melts containing oxide ions which would poise the membrane potentials.

For molten sodium fluoride, a sodium fluoride (5 wt %) *or* nickel fluoride (2–3 wt %)-doped hot-pressed boron nitride pellet was used as the diaphragm.[110] Presumably this functioned as a true salt-bridge of the type first used in aqueous solutions many years ago. The best results were obtained with sodium fluoride doping. The design of a reference electrode employing this principle is shown in Figure 3.

Another type of diaphragm is a lanthanum fluoride single crystal, which was introduced by Bronstein *et al.*[111] This was also used by Mamantov[112] in FLINAK and gave good results at 500°C, although some electronic conduction has been observed. These diaphragms function through the solid state mobility of the F^- ion and in principle their (minor) dissolution in the fluoride melts is inconsequential as the La^{3+} ion is likely to have a very cathodic deposition potential and therefore should not interfere with other electrode processes present. The single crystal shows good corrosion resistance in, e.g., sodium fluoride, but it dissolves very rapidly in cryolite melts.[110]

In this section, only a broad survey of the reversible electrodes available for emf measurements in melts at high temperatures has been given. However, a knowledge of the principles on which the many electrodes which have been devised are based should facilitate the choice of electrode for a particular experimental situation and help in the design of new electrodes if this is necessary because electrodes are not already available or fully developed.

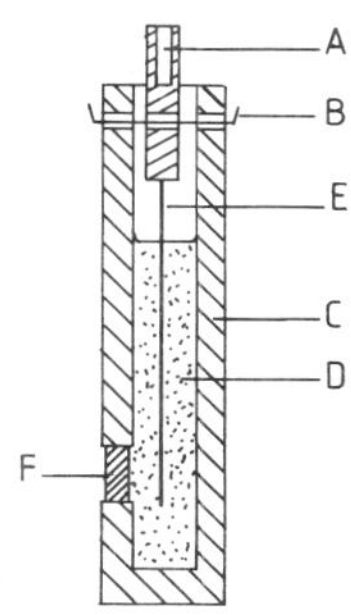

FIGURE 3. A reference electrode for use in fluoride melts. A: threaded nickel support; B: platinum wire supporting the compartment; C: boron nitride compartment; D: reference melt; E: spec. pure nickel wire; F: doped boron nitride pellet.

3. Atmosphere Control in Electrochemical Cells

3.1. Introduction

Little or no progress was made on the study of fused salts until measurements were made in a controlled atmosphere. Indeed many gaseous compounds of the external atmosphere react with the melt or with the cell materials, and it is often important to perform the experiments under an inert atmosphere. A suitable atmosphere can also play an active part in the experimental process in preventing the decomposition of the salt or in lowering the corrosion rate of the cell materials. Thus we shall first examine ways of reducing the level of the deleterious components of the air which can be present (air leakage or degassing process) in an inert atmosphere such as nitrogen or rare gases, and secondly we shall examine the handling of more specific atmospheres.

3.2. Influence of Some Air Components on Fused Salt Stability

At high temperatures *oxygen* is very aggressive and leads to the formation of metal oxides; for example, the activity of copper–gold electrodes changes continuously in LiCl–KCl baths when traces of oxygen (a few ppm) are present in the cell;[113] this effect is attributed to the selective oxidation of copper. An oxygen-free atmosphere is also required when fused alkali or alkaline earth metals are involved in the electrochemical reactions. It has been shown that the self-discharge of the LiAl/FeS battery becomes important under an oxygen atmosphere; this is attributed to the electronic conductivity of the melt caused by formation of Li_2O_2.[114] When carbon electrodes are used, as is often the case in the electrochemistry of fused salts, it is also important to be careful about the presence of oxygen which may first change the surface properties of the electrodes and finally destroy them. Exclusion of oxygen is essential when fused organic salts are used.[115]

Moisture has a deleterious effect in many fused salt experiments. Many compounds are highly sensitive to humidity, e.g., sulfides, nitrides, dissolved $AlCl_3$, chloroaluminates, fused metals, organic salts, etc., and even with alkali halides it can hydrolyze to give the nonvolatile O^{2-} or OH^- ions which may attack glass crucibles.[7] This contamination is very deleterious in the preparation of refractory metals by fused salt electrolysis: the presence of oxides leads to brittle metal.[116] Moreover, it has been shown[117] that at high temperatures the reactions $C + 2H_2O \rightarrow CO_2 + 2H_2$ or $C + H_2O \rightarrow CO + H_2$ affect the electrochemical properties of the carbon electrodes, an effect which is attributed to the presence of C–OH complexes.[118] Dry atmospheres are required when corrosive gases such as Cl_2, HF, and F_2 are used, since stainless steel pipes are rapidly corroded in the presence of moisture and it has been pointed out that condensation of HF–H_2O in copper tubing dissolves the protective copper fluoride.[119]

The presence of *carbon dioxide* should be avoided in baths having a basic character such as fused hydroxides[120] or dissolved oxides[121]; besides it decomposes the alkali sulfides. At high temperatures the attack of iron sulfide has been reported,[122] and many metals are oxidized. It should also be borne in mind that above 400°C the reaction $CO_2 + C \rightarrow 2\ CO$ occurs and can damage the carbon crucibles or electrodes.

In most cases *nitrogen* may be considered an inert gas. However, at high temperatures many metals give rise to the formation of nitrides. During the electrolysis of fused alkali chlorides (for the preparation of alkali metals) it has been shown that under nitrogen atmospheres, nitride ions are present in the bath and by anodic oxidation they give nitrogen oxide which is found in the chlorine.[123] Recently in fused salt batteries with a Li/B alloy anode, the formation of Li_3N and BN has been reported when an atmosphere containing nitrogen is used.[124]

Organic vapors may also be present in the cell atmosphere. Their decomposition at high temperatures gives carbon residues which pollute the bath. They may also react with the products of the electrochemical process such as alkali metals.

3.3. *Purification of an Inert Atmosphere*

This atmosphere generally consists of pure nitrogen, argon, or sometimes helium. Now that bottles of high purity gases are commercially available, the main problem is to avoid contamination by the ambient air. It is essential to use a vacuum-tight cell, as the pressure of the impurities coming from the air will be the leak rate of the apparatus. In this connection the gas circuit has to be carefully checked; for example, air leakages can occur in the pressure regulators. In our laboratory a simple preliminary test consists of placing the whole gas circuit and the cell under vacuum (10^{-2} torr) without pumping; the pressure variations are followed and a stable pressure of 10^{-1} torr, or less, during about 15 min is considered suitable for most of the experiments. To meet this requirement it is necessary to avoid rubber or polyethylene tubing; copper is a suitable material for this purpose and is easy to use, and glass-metal fittings are available. Even when a pipe contains a gas under a pressure greater than the atmospheric pressure the ambient air can diffuse through the pipe; this diffusion, called back-diffusion, is under the control of the partial pressure of the gases on each side of the tube. These permanent pollutants depend on the gas flow through the pipe; as an example the pollution by the back-diffusion of oxygen is given for different materials in Table II.[126]

Porous materials such as activated alumina and charcoal contain considerable quantities of gases which must be taken into consideration when the apparatus containing these materials has to be evacuated.

However, notwithstanding these precautions, it is worthwhile to introduce a simple gas cycle for the removal of the most noxious impurities. A very common

TABLE II. Oxygen Pollution (From the Air by Back Diffusion) of Pure Argon under a Pressure of 10 bar, Length of the Tube 1 m; Diameter 6.1 mm; Flow Rate of Argon 5 l h^{-1}

Materials	Amount of oxygen (ppm)
Mylar ($C_{10}H_8O_4$)	0.02
Nylon 6	0.05
Kel-F (Polytrifluorochloroethylene)	0.6
Perbunan	5.3
Neoprene	6.9
Polyethylene	11
Teflon $(CF_2)_x$	13
Polyvinyl	27
Natural rubber	40
Stainless steel	0
Copper	0
Metallic tubing with a 20 μm pore	20

problem is *water vapor* which can be removed by a drying agent which is introduced in a glass column. The relevant properties of a few solid drying agents are given in Table III.

Magnesium perchlorate: very efficient, high capacity, it may absorb 35 wt % of water. Its regeneration is quite difficult (48 h under vacuum at 240°C). With organic vapors compounds may be formed which explode when heated.

Molecular sieves[127]: they are synthetic zeolites (alumina + silicates). Their crystalline structure consists of holes with porous walls, the pore size being quite homogeneous: 3, 5, or 10 Å. The sieve absorbs only molecules of which the diameter is smaller than the pores. Polar molecules such as H_2O, CO_2, H_2S are strongly adsorbed. For example, with molecular sieves of 5 Å it is possible to reduce the water vapor and carbon dioxide content to 1 ppm. They also absorb many organic vapors. Molecular sieves of 3 Å do not absorb CO_2 and thus can be used to produce dry CO_2. An advantage of molecular sieves is that they are easily regenerated by heating under vacuum at about 320°C and that they will undergo almost an unlimited number of drying cycles.

Phosphorus pentoxide: very efficient, high capacity, but it may give off corrosive phosphoric acid. It cannot be regenerated.

Silica gel: its efficiency is similar to that of molecular sieves, its dehydrating properties being due to the same mechanism. It absorbs organic vapors and gases with higher boiling points. It may be regenerated at lower temperatures (12 h at 150°C under vacuum).

Activated alumina: has a greater efficiency than silica gel and a lower capacity. It may be rapidly regenerated by heating for a few hours under vacuum at 200°C.

Calcium chloride: is an anhydrous granular substance which has a low capacity. It may be used with many gases, becomes wet when exhausted, and its regeneration is impracticable.

Quicklime: its efficiency and capacity are not very high, it absorbs all acid gases. It is the only common drier suitable for ammonia.

A very efficient technique to eliminate certain impurities is to use a *cold trap* (see below) but it is necessary to have a ready supply of the coolant (solid CO_2 or liquid nitrogen).

Small proportions of *oxygen* can be removed by treatment with a *heated metal* such as lead or copper in the form of fine particles.[128] Oxygen is removed from nitrogen by passing the gas over *uranium nitride* at 500°–600°;[129] for permanent gases *titanium sponges* at 700°C have been used.[130] As pointed out above, *carbon dioxide* is in general removed by passing the gas through columns filled with *molecular sieves* (5 Å) or with pellets of *sodium hydroxide*. A cold trap with liquid nitrogen is also a very efficient way of eliminating CO_2.

Organic vapors can be eliminated on a bed of *copper chips* at high temperature (650°C). Bubbling through a sulfuric acid solution of K_2CrO_4 followed by an *Ascarite* (NaOH) filter is also very efficient.

When a *nitrogen-free* atmosphere is required this contaminant can be gettered (together with oxygen and carbon dioxide) with *heated* (at 650°C) *zirconium* or *titanium* (at 850°C).[131] The gas passes into a fused silica tube whose central portion is filled with chips or powdered metal. Water vapor and hydrocarbons tend also to react and release hydrogen, the only active gas likely to remain unabsorbed in the purification train.

If it is suspected that *solid suspensions* (dust) are present in the gas, they can be removed by passing them through a porous diaphragm. A *glass wool* plug may be used, followed by a *fritted glass disc*. When fine porosities are used the pressure drop across the filter is important. For example, with a com-

TABLE III. Solid Drying Agents[125]

	Efficiency[a]	Relative capacity[b]
	mg/l	*l*
Magnesium perchlorate [$Mg(C1O_4)_2$]	0.0002	1100
Activated alumina [Al_2O_3]	0.003	260
Molecular sieves 5 Å	0.004	220
Phosphorous pentoxide [P_2O_5]	0.004	570
Silica gel	0.070	320
Quicklime [CaO]	0.660	50
Sodium hydroxide [NaOH]	0.51	180
Calcium chloride [$CaCl_2$]	0.14	31

[a] For a flow of 225 ml/mm of nitrogen passing through 3 drying columns (I.D 14 mm, length 150 mm).
[b] For a given volume (unspecified) of drying agent.

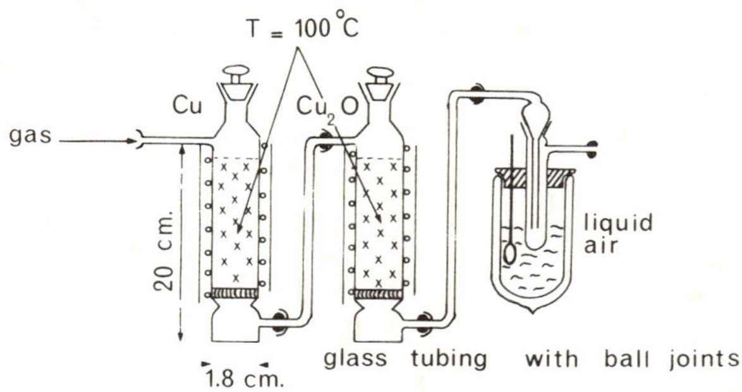

FIGURE 4. Simple gas train for the purification of an inert atmosphere.

mercial Pyrex fritted disc (thickness 3 mm, diameter 2 cm, pore size 5 μm), a nitrogen flow of 160ml/min requires a pressure difference of 50 torr.

A simple method of inert gas purification may be used to remove moisture, oxygen, carbon dioxide, and hydrocarbons. The principle is to reduce oxygen on copper and to oxidize organic substances on cuprous oxide (Cu_2O); the water vapor and carbon dioxide are then removed in a cold trap (Figure 4). Efficiency is greatly improved by the use of copper; in fine powder form, deposited on a diatomaceous earth, this catalyst gives high reaction rates at low temperatures (100°C) where the thermodynamics of reactions are favorable. Details of the preparation are given in Ref. 132. The color change from black for reduced copper, to green for Cu_2O or dark grey for CuO provides a useful hint that the limit is not exceeded.

When argon is used as the inert gas and liquid nitrogen as the coolant it is necessary that the liquid nitrogen level remains at all times below the trap to prevent the argon from liquifying. A simple glass bubble float indicator was found helpful for this purpose.[132] It is also essential to control the escape of the gas from the cell by bubbling it through vacuum oil or sulfuric acid. The system should be kept under a small positive pressure to prevent back-diffusion. With this method the oxygen pressure actually obtained must be less than 5.10^{-7} atm; with an assumed trap temperature of 100 K, the final vapor pressure of water should be approximately 10^{-15} atm and of carbon dioxide about 2.10^{-7} atm. In fact, the level of impurities is controlled by the quality of the gas connection. To obtain high performance it is essential that the system be tested and degassed under oil pump vacuum. A flexible connection between the gas train and the cell may be obtained by using glass tubing and ball joints. The system has been found to be extremely successful and reliable over a long period of time; the catalyst may be easily regenerated by hydrogen (first tower) or slowly oxidized to Cu_2O (second tower).

A gettering furnace capable of removing oxygen and water from inert gases is now commercially available.[133] It is capable of reducing 100 ppm O_2 and 100 ppm H_2O to a level of 0.1 ppm of each; its capacity is sufficient to purify about forty 10-m^3 cylinders containing a nominal impurity level of 10 ppm oxygen and water.

3.4. *Treatment of Specific Atmospheres*

Most of the experimental work in molten salt media is characterized by measurements at a somewhat high temperature (ranging from 200° to 1500°C). In these conditions the ionic liquids are quite aggressive, but the corrosion problem may be partly solved by maintaining the cell under a suitable atmosphere, and carefully purifying the melt components (for example, removal of traces of oxides which readily attack many materials). This may be carried out by treatment with reactive gases. The following is a discussion of the purification processes of the most common atmospheres involved in the handling of typical molten salts (chlorides, fluorides, nitrates, carbonates, and sulfates).

Hydrogen chloride: *Treatment of molten chlorides,*[134] *chloroaluminates.*[135]
Purified hydrogen chloride may be obtained by scrubbing the commercial gas with sulfuric acid which absorbs sulfur dioxide and water vapor. Final traces of moisture are removed by passing the gas through a mass of granular aluminum chloride.[136]

Chlorine: *Treatment of molten chlorides,*[137] *reference electrodes.*[138]
The impurities actually found in the commercial gas, totalling about 0.2%, are hexachloroethane, carbon tetrachloride, chloroform, bromine; also, traces of nitrogen monoxide have been detected in chlorine prepared by the electrolysis of fused alkali chlorides.[123] Distillation ($\Theta_E = -34.6°C$) removes the most common impurities (details of a laboratory column for this purification are given in Ref. (139)). In many experiments, this gas is only treated by concentrated H_2SO_4 and anhydrous $Mg(ClO_4)_2$.

Bromine: *Treatment of bromides.*[140]
The bromine vapors must be dried on a P_2O_5 column. A continuous distillation ($\Theta_E = 58.8°C$) device has been described[141] to operate in countercurrent electromigration processes.

Hydrogen fluoride: *Treatment of the fluoride baths which are proposed for use in nuclear reactors.*[142] *Preparation of acidic salts for fluorine production.*[143]
Anhydrous acid can be obtained by fractional distillation ($\Theta_E = 19.54°C$) as described in Ref.

(144). As regards the gas supply, the low volatility of the liquid is such that the temperature of the cylinder must be raised to about 35°C (temperatures greater than 45°C are dangerous). Hydrogen fluoride is very hygroscopic. The electrical conductivity of liquid hydrogen fluoride is very greatly influenced by traces of water and other impurities; this measurement constitutes a sensitive test of the purity of the product (at − 15°C the best value is $14.10^{-6}\ \Omega^{-1}\ cm^{-1}$).[144] The introduction of gaseous HF in a metallic apparatus (copper, Monel*) generates some traces of hydrogen at the beginning of the preparation; they may be removed by condensing HF.

Fluorine: *Produced by electrolysis of the salt 2HF–KF.*[145,146] The purification of fluorine from the oxygen it contains is carried out by selective adsorption on active charcoal,[147] by passing through a nickel tube at 300°C, and by distillation ($\Theta_E = -188.0$°C) at liquid nitrogen temperature. When HF is the only contaminant, it is possible to reduce the content to a fraction of 1% by passing over sodium fluoride. As with HF, adequate safety and first-aid precautions are essential.

Hydrogen: *Atmosphere for molten fluorides,*[119] *determination of the standard potential of the $HCl/H_2(g)$ couple.*[148] The most common impurities (oxygen and CO_2) are removed by passing the gas over copper chips at 650°C, followed by a column filled with $Mg(ClO_4)_2$ or by a liquid nitrogen trap. Nitrogen oxide gives nitrogen which in general may be used with the gas. To obtain very pure hydrogen, the best method is to diffuse the gas through a palladium tube[149] at 300°C, but the diffusion is quite slow; an alternative is provided by the use of thin-walled nickel tubing at 500°C (hydrogen diffuses at least a thousand times faster than other gases).

Carbon dioxide: *For use in carbonate melts*[150] *and fuel cells.*[151,152] The main impurities in cylinder gas are O_2, N_2, CO, H_2S, SO_2. Melville and Gowenlock[149] propose the following purification gas train: titanous chloride solution, sodium bicarbonate, copper sulfate,

* Monel is a nickel-based alloy containing (in w/o) 28–30% copper, 2–4% iron, and small amounts of manganese, silicon, and aluminum.

concentrated sulfuric acid, hot copper (700°C), and hot copper oxide (700°C).

Carbon monoxide: *For use as fuel for molten carbonate cells (with* CO_2[153] *and* H_2[154]*).*
A liquid air trap removes CO_2 and H_2O and saturated hydrocarbons; traces of iron carbonyl may be removed by passing through pre-ignited asbestos at 500°C.

Nitrogen dioxide: *Reaction studies in nitrite and nitrate melts.*[155,156]
The NO_2 is freed from oxygen by repeated fractionation (Θ_E = 21.1°C, equilibrium mixture NO_2–N_2O_4) between liquid nitrogen traps, the gas passing over P_2O_5.[149] NO_2 readily attacks most tap greases;* rubber and copper or copper alloy connections are not admissible. Calcium chloride is not a suitable drying agent as it may react to produce nitrosyl chloride.

Dinitrogen pentoxide: *Treatment of nitrate melts.*[157]
An atmosphere of N_2O_5 or, more exactly, of the mixture ($2NO_2 + 1/2\ O_2$) has been used to avoid nitrate decomposition and the attack of the Pyrex crucible which occurs in an inert atmosphere. It is prepared by adding P_2O_5 to frozen pure nitric acid;† the product is condensed in a trap at solid carbon dioxide temperature. The crystals are then slowly sublimed in an inert gas; the vapor pressure of N_2O_5 is 279 torr at 20°C. As with NO_2, it is important to insure that N_2O_5 does not come in contact with organic materials.

Sulfur trioxide: *Reaction studies in sulfate baths.*[155]
The sulfur trioxide prepared by dehydrating fuming sulfuric acid is purified by repeated distillation (Θ_E = 44.5°) and warming for six or seven hours over phosphorus pentoxide to remove the hydrate. Sulfur trioxide is available commercially; the vapor can be dehydrated by passing over phosphorus pentoxide.

Oxygen: *Oxygen electrodes*[158–161] *and use in gas mixtures for oxygenated salts:* $CO_2 + O_2$ *for carbonates,*[150,162] $NO_2 + O_2$ *for nitrates,*[163,164] $SO_3 + O_2$ *for*

* Phosphoric acid may be employed for all taps and ground joints.
† "Fuming" nitric acid can be used.

sulfates.[155] Traces of hydrogen are removed on platinized or palladized asbestos at 300°C; the resulting moisture is fixed on $Mg(ClO_4)_2$. The gas also contains traces of CO_2 which may be absorbed on soda lime.

3.5. Pressure Measurement

Most of the experiments are carried out under atmospheric pressure. As pointed out above, a slight positive pressure is maintained in the cell to prevent air getting in. This pressure arises from the bubbling of the escaping gas in a suitable liquid such as oil or sulfuric acid. Sometimes higher pressures are required to avoid the evaporation of a volatile constituent; the development of sodium[165] or lithium[114,165]/sulfur batteries requires the use of sealed cells, the vapor pressure of sulfur at 450°C being 890 torr.

The solubility of reactive gases is an important factor in the study of the electrochemical cells; many vapor pressure measurements have been carried out to determine the pressure dependence of the gas solubility.[166] Equilibration with known water vapor pressures was employed to adjust the water concentration in $LiNO_3$–KNO_3 eutectic.[167] Henry's law is adopted for calculating the precise concentration of water dissolved in the melt. Pressure changes can be used to measure the actual number of moles of gaseous reactant taken up or released by the melt when the volume of the cell is known; this determination has been carried out for example to study the equilibrium $Na_2O + 1/2\ O_2 \rightleftharpoons Na_2O_2$ in the electrochemical study of the system Na_2O–NaCl.[168]

Vapor pressures of a volatile compound of the melt can be measured by the transpiration method.[169] The amount of this component transported by a known flow of eluent inert gas is carefully determined after condensation in a weighing tube. Water vapor pressures of aqueous melts have been measured by a differential transpiration method.[170] The inert gas is first equilibrated with pure water at a known temperature and then successively passed through weighed drying tubes, equilibrated with the aqueous melt at a known temperature, and finally passed through a second set of drying tubes. The transpiration method does not yield directly the manometric pressure, but the true vapor pressure can be calculated if the equation of state of the vapor-carrier gas is known. For the cited example,[170] the vapor pressure varied from 47 to 338 torr; the correction to Dalton's law vapor pressure was never greater than 0.3 torr. The estimated maximum experimental error was ± 1 torr.

It is sometimes useful to look at the pressure influence on the electrochemical kinetics as in the case when gaseous products are involved in the electrochemical reaction. For example, it has been shown that the anodic oxidation of carbon electrodes in baths containing oxide ions is very sensitive to pressure variation,[171,172] a phenomenon which is attributed to the gas adsorption and to the gas coverage of the electrode. In the range 0–2 atm the pressure is measured on a mercury manometer with a cathetometer; the pressure depression is obtained by a rotary vacuum pump protected by a liquid nitrogen trap, which is required to remain just above the vapor pressure of the electrolyte.

Variable pressures are also used with gas electrodes to determine the pressure dependence of their emfs.[151,173] In this case, the partial pressure of the active gas is varied by use of the dilution method. For example, different N_2–Ar mixtures have been prepared in order to verify the thermodynamic behavior of the nitride–nitrogen electrode in LiCl–LiF mixtures.[174] Mechanical manometers such as membrane manometers or Bourdon gauges are useful to operate in a very large range of pressures.[175] The high-pressure field has been investigated mainly to determine the physicochemical properties of fused salts at constant volume[176] when the temperature is varied, but when the pressure increases the gas solubility becomes important and a cell without atmosphere has to be used.[177]

With corrosive gases a special diaphragm attachment can be used: one side of the diaphragm is exposed to the system and the other side is filled with a liquid which transmits the pressure to the Bourdon tube.

3.6. *Measurement of Gaseous Flow Rate*

In many experiments an exact determination of the gaseous flow rate is not required, it is only estimated from the bubble rate in a liquid bubbler at the gas exit of the cell; however, low cost meters now exist which give an accurate determination of the gaseous flow rate at the laboratory scale.[133]

3.6.1. *Ball Flowmeter*

The level of a ball in a vertical metering tube of standard tapered bore is proportional to the flow rate. The gas flows upward through the tube and raises the ball, increasing the annular area between the tube and the ball until the upward force is equal to the weight of the float. The stabilization of the ball is insured by longitudinal grooves. Balls of different materials such as glass, sapphire, stainless steel, and tantalum may be used simultaneously. These flowmeters are available for flows as low as 4 ml and as high as 60,000 ml per min. An accuracy of 3% of full scale may be obtained. To obtain significant and reproducible results it is necessary to remove dust and spray (oil, sulphuric acid, etc.) from the gas. More reproducible results may be obtained with a free rotating float which operates without friction.

The calibration of the flowmeter is usually carried out by timing the gas discharge through a volumetric displacement flask.[178] The vapor pressure of the fluid used must be taken into account in the gas volume correction. It often happens that the flowmeter is used with another gas B different from the calibration gas A; when the specific masses ρ of the two gases are known, the flow rates Q_A and Q_B are related by the formula[125]:

$$Q_B = Q_A\left(\frac{\rho_A}{\rho_B}\right)^{1/2}$$

which may be used to modify the calibration curve. Ball flowmeters are affected by pressure and temperature variations; it is therefore always advisable to control the pressure with a good regulator. When the tube is equipped with a control valve at the inlet, the correct readings require a constant outlet pressure close to atmospheric. When the downstream pressure may vary, the control valve must be located at the outlet of the tube, the gas pressure being adjusted to the pressure for which the tube was calibrated.[133]

3.6.2. Capillary Flowmeters

When the gas passes through a capillary tube the resulting pressure drop is related to the flow rate Q of the gas. For viscous flow (low velocities) Q is given by Poiseuille's law.[125]

$$Q = \frac{(P_1 - P_2)\left(1 + \frac{P_1 - P_2}{2P_2}\right)}{r}$$

where $r = 128\eta L/\pi d^4$; Q = flow rate (cm^3/s) at the outlet of the capillary; P_1 = pressure upstream ($dyne/cm^2$); P_2 = pressure downstream ($dyne/cm^2$); L = length of the capillary (cm); d = diameter of the capillary (cm); and η = viscosity of the gas (poise).

The pressure difference is easily measured by introducing the capillary between the two branches of a manometer filled with a suitable liquid such as mercury or a low-vapor-pressure liquid (silicone oil); however, the solubility of the gas in this liquid must also be considered. On this principle many arrangements have been described.[175] For example, Burton and Nichol[179] have designed a microflowmeter which is sensitive to flow as low as 0.01 ml per minute. With suitable materials these meters can be used with corrosive gases. Copper tubing capillary flowmeters were employed for metering the rate of fluorine and hydrogen fluoride.[180,181]

The previous meters give a direct reading of the flow rate. However, it is often useful for the control and the optimization of the electrochemical cell to obtain the flow rate of gases as an electric signal. This can be achieved by use of the following flowmeters:

3.6.3. Turbine Flowmeters

In this apparatus, the gas passes axially through a rotor which is mounted between low-friction bearings. In most commercially available instruments the rotor is made of magnetic stainless steel. Variations in magnetic field near the rotor are used to determine its speed. Light free-running rotors have been designed for a miniature turbine flowmeter, the speed being determined by a photoelectric method.[182] This apparatus provides a rapid response to a change in flow rate, and measures gaseous flow rate down to 1.8 liter/mn.

3.6.4. *Thermal Flowmeters*[183]

They are based on the rate of cooling or on the heat transfer from a heated element placed in the gaseous stream. Three main types are used: *hot wires,** *thermistors,†* and *thermocouples.‡*

**Hot wires:* A heating current is passed through a sensitive wire (generally made of platinum, platinum alloy, or tungsten), then the wire is heated by the Joule effect at a temperature higher than that of the flowing medium. The heat loss depends on the velocity of the gas or, more exactly, on the mass flow $Q = \rho AU$ (A = area of the tube section; ρ = specific mass of the gas; U = velocity of the gas). An equilibrium resistance of the wire corresponding to the temperature change is used to measure the gas flow. To compensate for the external temperature change a differential measurement is carried out by making an adjacent arm of the Wheatstone bridge out of the same materials as the element and immersing it in the fluid; the heating in this compensating arm must be negligible. With this technique, a very large range of flow measurements may be covered from 0.1 ml mn^{-1} up to 10 m^3 mn^{-1}. In general, the accuracy is about $\pm 0.5\%$ full scale for a constant temperature and pressure; changes of temperature of $\pm 30°C$ and of pressure from 0.3 to 3 atm of the flowing gas do not change the indications by more than $\pm 2\%$. Dirt and smoke particles in the atmosphere can cause serious changes in the calibration of the heat probe.

†*Thermistor flowmeters* work on the same principle; the only fundamental difference is the increase of the resistance when the temperature decreases, the reverse of the effect observed with hot wires. The advantages of the thermistors are that the temperature coefficient of resistance is very large and that a large resistance can be obtained with a quite small element.[184] Because of this great sensitivity, the working temperature cannot be very high (60°–70°C above the fluid temperature).

‡*Heated thermocouples* can also be used to determine the heat loss due to the gas flow. To compensate for changes in the fluid temperature, two thermocouples are immersed in the fluid; one is heated and the emf difference gives a sensitive measure of the flow of the gas. The accuracy of routine apparatus is about $\pm 2\%$ full scale. Variations of pressure of $\pm 50\%$ between 0 and 40°C do not change the calibration.

Electronic units which sense the mass flow of gases by measuring the differential cooling between two points are still quite expensive. They give measurements that are dependent on the heat capacity of the gas, but they provide accurate readings regardless of temperature and pressure changes in the gas stream.

3.7. Regulation of the Gas Flow

The usual needle valves operate satisfactorily for most experiments, however, the very narrow opening which forms the flow resistance is subject to slight deformation or eccentricity of the needle. More reliable valves may be obtained using a capillary regulating valve;[185] however, these systems depend on the stability of the pressure regulators.

When it is necessary to provide a constant flow rate of gas in the presence of pressure change in the cell, a compensating device is required. This may be achieved by incorporating a blowoff tube in the circuit.[151,186]

When electronic flowmeters are used, a valve controlled by the flow transducer may be used automatically to compensate the pressure variations and to maintain a constant stream.[133]

3.8. Gas Mixtures

3.8.1. Discontinuous Preparation

The method most often used is to mix the gases at constant pressure and then calculate the composition from the observed volume increase. This can be carried out in a calibrated flask filled with a low-vapor-pressure liquid (mercury, vacuum oil, saturated-salt solution); the lower stopcock of the flask is connected to a levelling bulb and the pressure may be adjusted by lowering the bulb. When a component is introduced into a vessel which already contains a gas it must be remembered that there is always the risk that some gas will escape while the next component is introduced. This risk can be reduced by using a large pressure difference.

3.8.2. Continuous Supply

For electrochemical cells a continuous supply of gas is generally required. When a gas mixture is used it can easily be prepared by adding the different components continuously after an accurate measurement of their flow rates. A suitable gas regulator can be constituted by a ball-flowmeter unit carefully calibrated for the pure components; high accuracy valves control the gas inlet. At the outlet the different gas flows are introduced in a mixing chamber to obtain a homogeneous mixture. An accuracy of $\pm 5\%$ can be achieved with this unit. The gas regulators are back-pressure compensated; correct readings are obtained regardless of the back-pressure provided that the inlet pressures are maintained at the levels for which the tubes were calibrated.[133]

A simple way of adding a certain amount of gas to a gas stream is to use controlled capillary leaks. A fine capillary using a needle valve in conjunction with a manometer gives a constant addition rate. Volumetric pumps are now available for the preparation of gas mixtures[187] and these piston pumps produce multicomponent mixtures using the dynamic injection method. The setting of the mixing ratio will be by switch gears (in a concentration range of 1–99%); the forwarding capacity is about 300 ml/mm up to 10 liter/mm and oil is used to lubricate the piston. The gas solubility in the oil must be considered (for example, sulfur dioxide is extremely soluble in silicone oil). Oil spray may also be formed by contact of the gases with the oil film in the cylinder, but it can be removed by using an activated carbon filter. By a careful choice of pump materials and of the oil, all gases may be mixed with this system except NO_2, Cl_2, and F_2.

3.8.3. Saturation-Vapor Pressure Method

Oxygen–nitrogen dioxide mixing can be carried out[188] by passing a stream of pure oxygen slowly through boiling liquid dinitrogen tetroxide (normal boiling point: 21°C). Then, by partial condensation at a suitable temperature T, the composition of the gas mixture (O_2–NO_2–N_2O_4) is calculated from the vapor pressure of liquid N_2O_4, the degree of dissociation of N_2O_4, and the total pressure. The temperature dependence of the vapor pressure of liquid sulfur trioxide was also used to prepare $O_2 - SO_3$ mixtures in nitrogen.[155] The flows of nitrogen and oxygen were controlled using pressure differential flowmeters with *n*-butyl phthalate in the manometers. This mixture was passed through liquid SO_3, (boiling point $\Theta_E = 44.5°C$) maintained at about 10°C hotter than the subsequent U-tube condensing column, and the temperature of the latter was varied to obtain the desired effluent rate of SO_3. Mixtures of O_2–H_2O (*g*) were prepared on the same principle by Zambonin[189] to study the influence of the partial pressure of water on the oxygen electrode in molten nitrates.

To obtain stable composition mixtures, it is necessary to use an apparatus consisting of inert materials. Rubber or plastic tubing are in general not admissible. For example, it has been indicated that Tygon* tubing considerably reduces the flow rate of gaseous chlorine[190] moreover, even with polyfluoropropylene tubings, an acceptable adsorption of hydrogen chloride occurs. In static conditions, we have observed a pressure drop of HCl of about 40 torr during 1 h in a Viton tubing of 4 mm i.d.

When a variation of the composition is not required, a premixed gas solution of known composition permits greater experimental accuracy. Several gas manufacturers specialize in the preparation of gas mixtures made to the specification of the buyer.

3.9. Gas Analysis

In an electrochemical cell the analysis of the gaseous compounds may be performed for two different purposes. The first one is to control the purity of the atmosphere and detect the amount of noxious components; the second is to

* Norton trade mark (polyvnyl chloride).

analyze the gaseous products of the electrochemical reaction. Apart from its interest in the control of industrial electrochemical cells, the gas analysis gives valuable information about the mechanism of the electrochemical reaction itself.

3.9.1. Chemical Reagents

The oldest technique to analyze a gas mixture is successively to remove the different components by using specific reagents. The pressure–volume change of the gas or the weight increase of the reagent tube measures the amount of the analyzed constituent. Following is a list of the gases analyzed, and the reagents which can be successively used for the atmosphere control.[149]

Gases	Reagents
H_2O	$KOH–P_2O_5$
CO_2, HCl or other acid gases	KOH
Oxygen	Phosphorus (white)
CO	Silver oxide, cupric oxide, KOH
Hydrogen	Cupric oxide

With a trained operator this technique gives accurate results (about $\pm 0.1\%$) but it remains quite time consuming; it does not provide direct readings and it is not convenient for a continuous control of the cell atmosphere; also it is not very specific. More efficient methods are now available such as mass spectrometry, infrared adsorption, and gas chromatography. For continuous routine analyses of binary mixtures, the measurement of thermal conductivity of the gas constitutes a convenient low-cost technique.

3.9.2. Mass Spectrometry[191]

This technique can be used for the analysis of a small quantity of gaseous samples.[121,192] It leads to a rapid identification of the different constituents of a mixture. Indeed, each pure component is characterized by its mass spectrum or "cracking pattern," which corresponds to the ionized fragments from gas molecules subjected to the electron bombardment. Either the molecule (or parent ion) or the most abundant ion fragment is used as the base peak. The principle of the analysis of gas mixtures is based on the following assumptions:

a) The base (or reference) peak is proportional to the partial pressure of the corresponding component in the mixture admitted to the ionization chamber;

b) A mixture spectrum results from a linear superposition of the mass spectra of the various components;

c) The cracking pattern characteristic of any particular molecular compound remains stable for a reasonably long period.

As differences in cracking pattern data for instruments of the same type are not very important, published data can be used to obtain rapid identification of

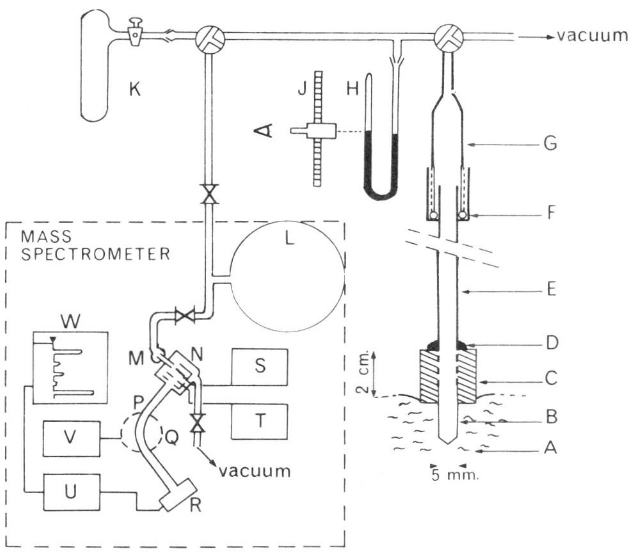

FIGURE 5. In situ analysis of the gaseous products on a graphite electrode. *A*: liquid salt; *B*: porous graphite—thickness of the wall: 0.5–1.0 mm; *C*: graphite block; *D*: silicate cement; *E*: stainless steel tubing; *F*: torion joint; *G*: Pyrex tubing; *H*: mercury manometer; *J*: cathetometer; *K*: sample tube; *L*: gas inlet reservoir; *M*: calibrated gas leak; *N*: ionization chamber; *P*: tube of the mass spectrometer; *Q*: electromagnet; *R*: ion collector; *S*: electron beam control unit; *T*: stabilized high voltage unit; *U*: amplifier; *V*: magnet current control; *W*: mass spectra recorder.

an unknown component.[193] However, for quantitative analytical work it is necessary to calibrate the apparatus first using pure samples of the various components under investigation. It is also of value to use internal reference standards such as neon and argon to give warning of significant instrumental changes which then require recalibration for all the pure components of the mixture being examined.

The analysis can be improved by a preliminary separation of some components by suitable techniques such as fractional distillation or gas chromatography.[194] A very high sensitivity can be achieved from the isotopic dilution technique[195] which is well adapted to the mass spectrometry analysis. In our laboratory the formation of nitrogen oxide during the electrolysis of fused alkali-chlorides has been studied by the use of NO labelled with nitrogen 15:[123] 10^{-7} mole of NO can be detected with an accuracy of $\pm$ 1%. It has been shown that the anodic chlorine contains a few hundred ppm of NO, but since NO is a chain inhibitor the chlorine produced in this way is unsuitable for photochemical reactions.

A direct mass spectrometric analysis of the gases produced by the electrochemical reaction has been described.[117] It concerns the electrolysis of alumina dissolved in a cryolite bath. As the electrolyte does not wet the graphite, it has been possible to use a hollow anode; the interior of the electrode is maintained at a low pressure and communicates with the sampling tube or directly with the mass spectrometer through a calibrated capillary (Figure 5). The influence of the electrochemical conditions on the gas composition can be instantaneously detected.

3.9.3. *Gas Chromatography*[196]

In gas chromatography, the separation of the components is effected by their partition between a stationary phase with a large surface and a moving phase (eluent). The success of gas chromatography is due to its simplicity and its high separation power. Gas chromatographs are relatively cheap, their maintenance costs are low, and their operation requires no special qualification. Very small samples of about 1 ml of gas are sufficient to obtain accurate analysis. With suitable apparatus the analysis of multicomponent systems can be performed in seconds. This technique is convenient for routine quantitative analysis of atmospheres the components of which are known. Errors due to the apparatus and to operation conditions may be considerably reduced by calibration with standard samples with known compositions.

In general, it is difficult to identify unequivocally the components of a sample containing unknown compounds. The corresponding peaks may be compared to the retention data of pure substances or model mixtures; each component is then identified from the increase of its peak after the addition of the pure substance to the sample. However, to identify the eluted components it is often necessary to use auxiliary methods such as mass spectrometry or infrared adsorption. The gas chromatography acts as a separation technique as does fractional distillation, but it provides a separation factor which may be varied between large limits by the choice of the stationary phase.

3.9.4. *Infrared Absorption*[197]

The infrared spectra of many gaseous substances are now well-established.[198] To obtain a spectrum small quantities of the substance (5–10 μg) are generally required, and the sample is not consumed. This technique furnishes useful information on the structure of the molecule and is of value in identifying unknown compounds.

As the intensity of the infrared spectrum depends on the partial pressure of the substance concerned, quantitative analysis can be carried out, but the results depend also on the total pressure and on the nature of the other components. Thus it is usually necessary to calibrate for each constituent of the mixture. The use of the infrared absorption method for gas analysis is limited by the overlapping absorption of different molecules, by the presence of unknown compounds, and by the magnitude of the molar extinction coefficient (the symmetric diatomic molecules such as O_2, N_2, Cl_2, etc., do not give an infrared spectrum).

Infrared spectroscopy is convenient for the continuous control of the gas evolved from an electrochemical reaction. It is commonly used to determine the composition of the gaseous effluent of cryolite cells (determination of the CO_2/CO ratio).[199] The purity of the fluorine produced by the electrolysis of the acid salt 2HF–KF is controlled by the infrared technique (detection of CF_4, HF, C_2F_4, CO_2, etc.,), but in this case the usual rock salt windows are not suitable; calcium or barium fluoride crystals have to be used.[200]

3.9.5. *Thermal Conductivity*[201]

The analysis of binary mixtures of gases of different thermal conductivity can be carried out by thermal conductivity gauges (katharometers) which measure the loss of heat from a hot body surrounded by the gas. The katharometer is essentially a simple instrument and given a supply of clean, dry gas will continue to operate almost indefinitely. However, it will analyze only essentially binary mixtures (H_2–N_2, CO–CO_2, Cl_2–Ar, N_2–O_2, etc.) and it is important to remove interfering components such as water vapor. The sensitivity depends on the nature of the two gases, a change of about 0.001% carbon dioxide or 0.0001% hydrogen in an inert gas atmosphere should be detectable. In many instruments, the sensitive chamber opens at one end and the gas changes by diffusion only; in this case, the readings are not dependent on the rate of gas flow, but their response to composition changes is quite slow (about 10 s). In other apparatuses the gas flows through the chamber and this construction gives a much higher speed of response (0.2 s).

3.9.6. *Miscellaneous*

An apparatus has been developed for the electrochemical determination of hydrogen in inert gas mixtures.[202] The cell involves hydrogen electrodes made of porous graphite activated with platinum black. This potentiometric method results in an accuracy to better than $\pm 0.2\%$ and is applicable for the continuous determinations of hydrogen varying from 100% to 0.1% in the inert gas mixture.

A vacuum ultraviolet absorption technique for the determination of trace concentrations of water in nitrogen, oxygen, or carbon dioxide has been proposed.[203]

Certain soluble gases can be determined by dissolution in an appropriate solution contained in a conductance cell; for example, CO_2 has been analyzed by absorbing the gas in a solution of barium hydroxide.[204] Traces of HCl are detected by changes of the conductivity of pure water.[149]

The use of gas-sensitive electrodes[205] can also be considered to determine the partial pressure of some soluble gases in water solutions. A device is described for continuous analysis of HF in air.[206,207]

4. The Electrochemical Cell

4.1. Introduction

Experimental measurements in the fused salt field are characterized by great difficulties which center upon the vulnerability to attack of container materials and the stability of the melt. These problems become predominant at high temperatures. In general, the cell atmosphere has to be carefully controlled which requires the use of a gas-tight apparatus. The cell design results from a compromise between the contraints arising from the aggressiveness of the melt and the necessity for easy handling of the electrodes and the electrolyte. Sampling of the reaction products is often desirable; moreover the ultimate purification of the salt is generally done in the cell itself to avoid contamination during the transfer operation.

This section deals with the realization of electrochemical cells; the system itself (electrolyte and electrodes) is not investigated. The most common cells are described and then special designs are given for some particular experiments.

4.2. Cell Designs

Three main parts are distinguished: 1) the salt container; 2) the inner materials: temperature measurements, crucible suspension, baffle, etc.; and 3) the outer jacket with the cell head. A typical arrangement is shown in Figure 6.

Sometimes, in order to remove solid suspensions which are present in the melt, a filtering device is included as part of the cell; two main arrangements can be used, an internal crucible with a fritted disc soldered at its bottom,[208] or an external container linked to the main crucible.[209] In the first arrangement, after the solid salt is introduced in the external crucible, the temperature is raised and the melt slowly filters into the fritted compartments maintained under vacuum (Figure 7). Sometimes, the salt is treated in a separate compartment and then pushed into the cell through a filtering pipe (Figure 8).[210]

In general, cylindrical crucibles are used; however when spectrophotometric studies are coupled with electrolysis planar windows are required and, in this case, square glass tubings may be employed.[211] For high temperatures (with silicate melts, for example) the cell may be a solid block of refractory material such as MgO in which two cavities are cut to contain the melt; the separation between the compartments should be as thin as possible to decrease the resistance of the cell[212] (Figure 9).

In general, dipping electrodes are used with vertical electrical connections. However, in some cases it may be useful to have an electrical connection through the walls of the container as in the case when a metallic layer constitutes the cathode at the bottom of the cell. This arrangement is found in large-scale cells with ceramic block linings.

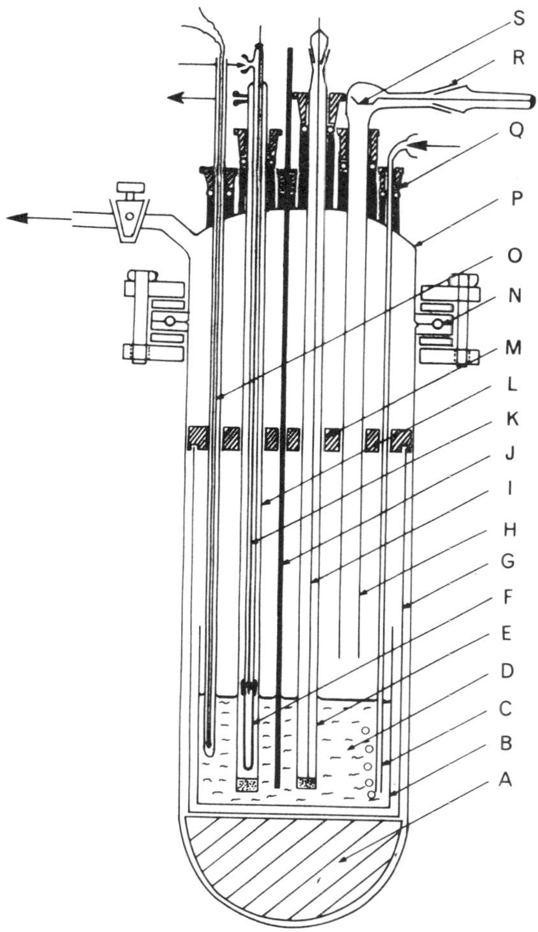

FIGURE 6. A typical electrochemical cell under a controlled atmosphere with an integrated feeding device. *A:* firebrick support; *B:* crucible; *C:* gas bubbling; *D:* fused electrolyte; *E:* tube with a fritted disc; *F:* counter-electrode (gas electrode); *G:* splash guard; *H:* feeding tube; *I:* reference electrode; *J:* working electrode; *K:* inner tube for gas inlet; *L:* tube with a fritted disc; *M:* separator and electrode guide; *N:* O-Ring; *O:* thermocouple; *P:* cell header; *Q:* tight Torion joint; *R:* ground point; *S:* rotating spoon for additional solute introduction.

In Figure 10 an arrangement is given for supplying the current at the bottom of a laboratory cell for the electrolysis of alumina dissolved in the cryolite bath. It is often of value to analyze the reaction products and, in this case, the working electrodes are separated from the other electrolytic compartments by a fritted glass or a sand bed. A chemical analysis of the dissolved products is done after

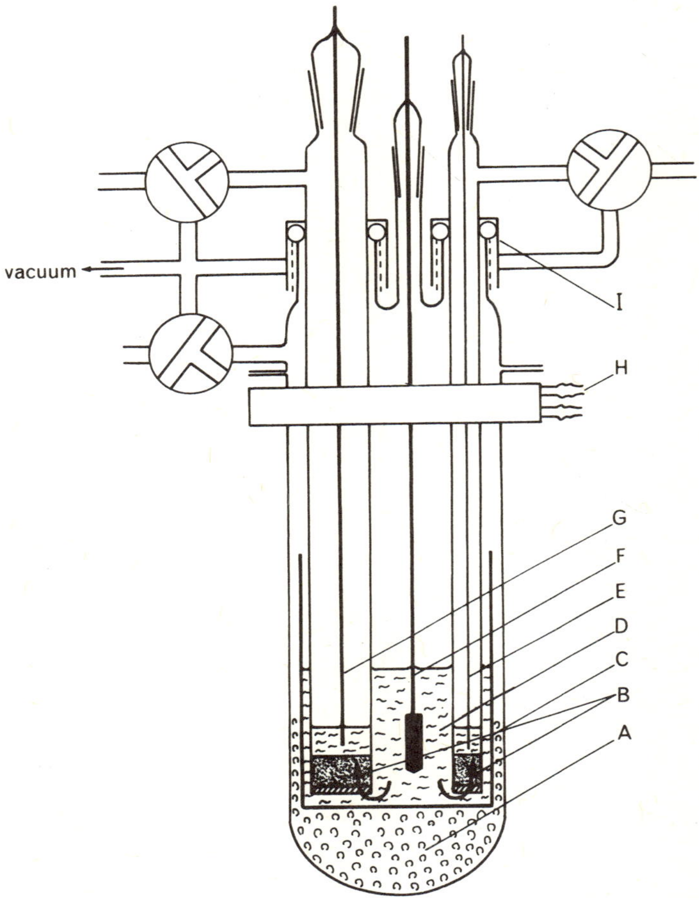

FIGURE 7. Electrochemical cell with an "in situ" filtering device for the electrolyte. The salt is fused in the crucible under a suitable atmosphere; then it is drawn into the two fritted compartments (working and reference). *A:* glass wool; *B:* fritted glass (n° 4 or 5—pore size 15–5 μm); *C:* crucible; *D:* electrolyte; *E:* reference electrode; *F:* counter-electrode; *G:* working electrode; *H:* cooling water; *I:* SVL joints.

sampling of the bath by means of aspiration through a dipper tube.[213] The gaseous products may be continuously analyzed by a direct sampling in the compartments, and an on-line analysis of the evolved gases has been proposed using a hollow electrode.[214]

Sometimes, to avoid the inconvenience of a large ohmic drop through a porous plug, the electrode compartment is surrounded by a bell to collect the gas or the liquid product (such as liquid metal in alkali metal or alkaline earth metal production.[213]

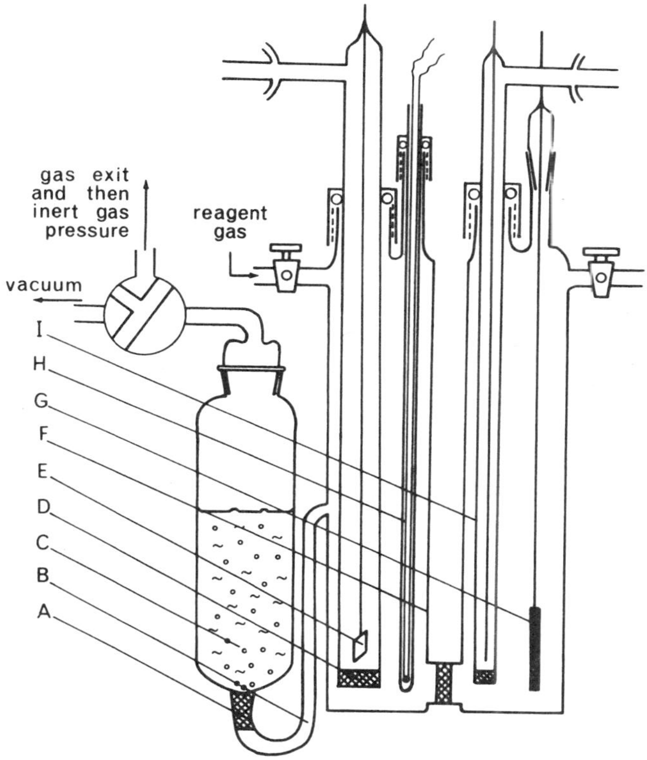

FIGURE 8. Cell with an auxiliary compartment for the salt preparation. In the actual cell the three compartments are placed around a central axis and introduced into a tubular furnace. The salt is dried under vacuum, treated by bubbling of a suitable atmosphere, and then pushed into the electrochemical cell (see Ref. 209). *A:* fritted tube; *B:* Pyrex tubing; *C:* purification compartment; *D:* fritted disc; *E:* working electrode; *F:* fritted disc; *G:* counter-electrode; *H:* thermocouple; *I:* reference compartment.

Changes in concentration of the electroactive species are often required for experimental purposes. These materials may be electrochemically generated but sometimes they are introduced by external additions. For gaseous compounds such as HF in the fluorine production,[215] $TiCl_4$ in the titanium deposition,[216] etc., bubbling is generally used to supply the feed material. The flow rate must be carefully adapted to the dissolution rate in order to avoid a depression in the inlet tube.

When solid materials have to be dissolved in the melt they must be intro-

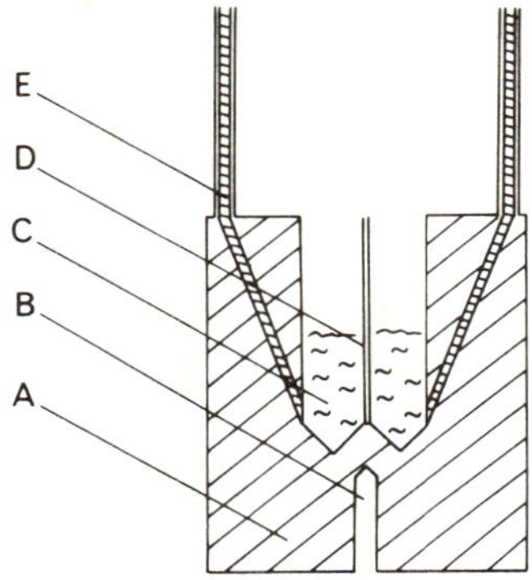

FIGURE 9. Cell for high-temperature experiments in silicate melts (see Ref. 212). *A:* solid block of MgO; *B:* thermocouple well; *C:* silicate melt; *D:* thin MgO diaphragm; *E:* Iron electrodes protected by a magnesia coating.

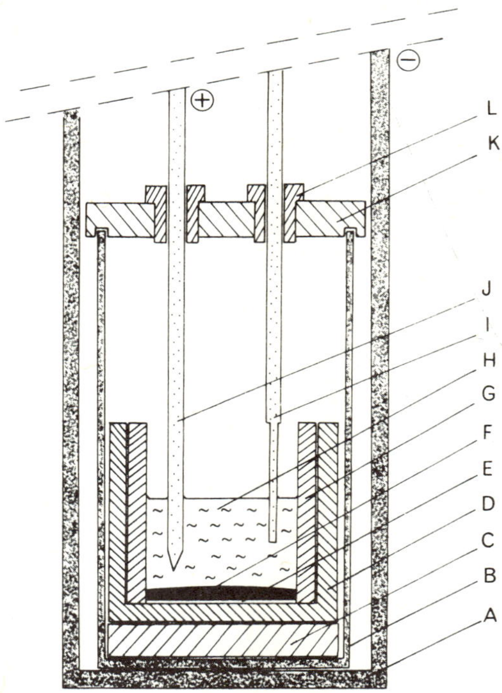

FIGURE 10. Laboratory cell for the aluminum preparation, the metallic layer of aluminum constitutes the cathode. *A:* stainless steel outer cell; *B:* stainless steel splash guard; *C:* piece of nickel; *D:* graphite crucible; *E:* titanium boride disc; *F:* liquid aluminum; *G:* boron nitride shielding; *H:* fused cryolite; *I:* nickel reference electrode; *J:* graphite anode; *K:* graphite cover; *L:* boron nitride insulators.

duced in the cell atmosphere at the beginning of the experiment as represented in Figure 6. Such methods have been used to study the influence of alumina concentration in the electrolytic preparation of aluminum; when γ-alumina is used it is important to keep it under a controlled atmosphere because it adsorbs many gaseous compounds such as water vapor and the presence of such impurities strongly affects the electrode response.[217]

The analysis of the electrodes themselves can be carried out during the experiment by use of a lock chamber.[213,218] A sliding compartment has been described by Chemla and Grinevitch[219] to obtain an easy change of the electrodes during the experiment (Figure 11).

In general, the cell is heated by an external furnace; the temperature gradient

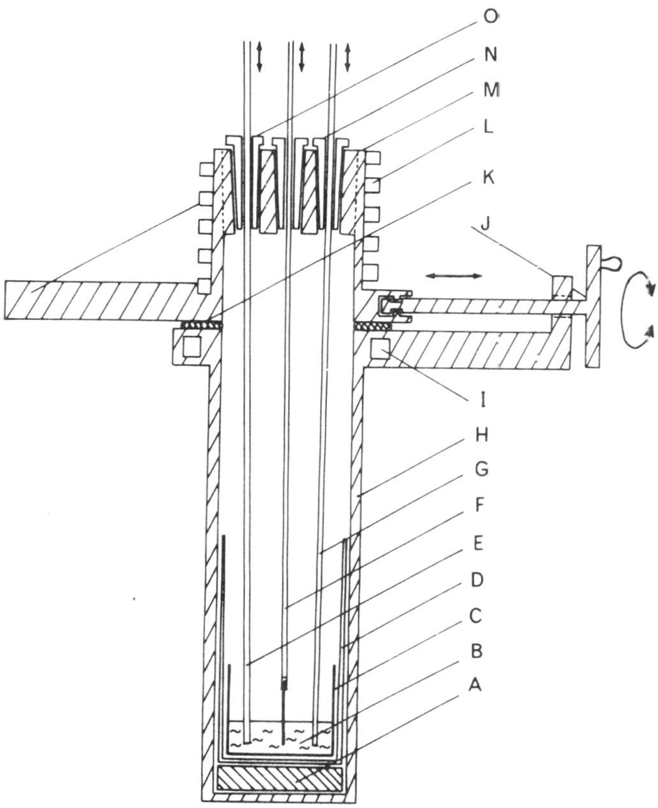

FIGURE 11. Electrochemical cell with a lock chamber to remove the electrodes during the experiment (see Ref. 219). *A:* nickel piece; *B:* fused electrolyte; *C:* vitreous carbon crucible; *D:* stainless steel splash guard; *E:* graphite counter-electrode; *F:* molybdenum reference electrode; *G:* metallic working electrode; *H:* stainless steel outer cell; *I:* cooling water; *J:* crank. *K:* brass cushion; *L:* cooling coil; *M:* bronze header; *N:* sliding cover plate; *O:* Teflon plug.

may be kept smoothed by a metallic tube in which there are observation ports. Indeed, visual observations and photographs of the electrodes are often required in order better to understand the electrode effects. In cryolite melts, cells with windows made of sapphire[220,221] and of quartz[221] have been tested. A diamond window cell for molten fluorides has been described.[222]

At low temperatures the corrosion problems are in general negligible and the salt can be directly introduced into the glass vessel (outer cell), and the crucible and the splash tube are no longer necessary (for organic salts for example).

4.3. Materials for Salt Containers

These materials must be completely inert to the fused salt to avoid the introduction of impurities such as dissolved metal or oxide ions, etc. Their lifetimes at the working temperature must be much longer than the duration of the experiment; this factor varies from a few hours for laboratory experiments to several months for industrial applications of fused salts,[223] such as nuclear reactors,[224] batteries,[225] and energy storage.[226]

Attack by a melt usually involves the penetration of solid containers by intergranular solution processes, and the rate of this attack depends primarily upon kinetic factors and thus upon the temperature.

When possible, the crucible is made of transparent material, visual observation bringing valuable information about changes on the electrode surface such as the nucleation process, gas bubble formation, electrode effect, etc.

To avoid secondary electrochemical reactions, it is sometimes necessary to use an electrical insulator material; however, metallic or graphite crucibles are commonly used and in many cases they are used as counter-electrodes.[227,228] This electrical polarization may be considered to reduce the corrosion.[227,228]

For a given electrolyte the nature of the suitable container depends on the experimental conditions: cell atmosphere, temperature, and the presence of dissolved materials or impurities. For example, the presence of moisture or oxide ions strongly enhances the attack of refractory materials (Al_2O_3, ZrO_2, BeO . . .) and the presence of traces of air in the cell atmosphere is incompatible with graphite or carbon crucibles; copper tubings are rapidly corroded by HF containing water vapour,[229] etc.

Specific corrosion problems are examined in the chapter dealing with fused electrolytes, but we summarize here the conclusions concerning the most commonly encountered electrolytes.

4.3.1. Fluorides

Fluoride melts are involved in three main research fields: the preparation of fluorine, the experiments related to the molten salt nuclear reactor, and the electrolytic preparation of aluminum.

4.3.2. Fluorine Preparation

The industrial electrolyte is the mixture 2HF–KF with which the working temperature is about 85°C[230] and cells are made of Nickel or of Monel.* However with these materials it has been shown that small amounts of nickel ions are present in the bath and lead to a metallic deposit on the cathode, and so these materials have to be protected by a cathodic polarization. During laboratory experiments, we have observed a nickel or a copper deposit on the platinum foil of the hydrogen reference electrode when Monel, nickel, or copper tubings are used;[146] this deposit appears even when the metallic tubings are not in contact with the bath and so all parts of these metallic pieces directly in contact with the hydrogen fluoride vapor have to be coated with Teflon. At low temperatures (below 90°C) cells made of Altuglass† are used at the laboratory scale. After some time, cracks appear, mainly above the level of the liquid, but such cells (cell wall : 1 cm, internal diam. : 10 cm) can operate for several weeks and can also tolerate freezing periods though this treatment greatly enhances their fragility. Cells made of iron have been described.[228] To avoid corrosion, the wall of the cell was used as the cathode and the bottom of the cell was covered with a Teflon plate. Transparent polytrifluorochloroethylene Kel-F also constitutes a suitable material.[228]

4.3.3. Nuclear Energy

The extensive research performed in order to build a molten salt nuclear reactor has given a decisive impulse to the corrosion studies in molten fluoride baths. A large part of the experiments is devoted to the mixture $LiF–BeF_2$, but the conclusions drawn from these studies are applicable to most of the alkali fluoride mixtures.

The purification of molten fluorides can be performed in graphite or in a limited number of metals such as nickel, copper, platinum, and gold. Nickel is the preferred choice for containing most molten fluoride systems; however, when available, the best choice is a nickel-based alloy such as Hastelloy N,‡ the composition of which has recently been improved to avoid cracking problems.[231]

With silica crucibles, which have been proposed as containers for fused fluorides, the following reaction takes place:

$$SiO_2 + 4\,F^- \rightleftharpoons SiF_4 + 2\,O^{2-}$$

* Monel K 500 antimagnetic, nickel base, copper 30%, aluminum 2.9%, iron 2%, manganese 1.5%, titanium 0.6%, silicon 0.5%, chromium 0.15%, sulfur 0.01%.

† Altuglass : polymethylmethacrylate; Trade Mark CdF chimie.

‡ Hastelloy N: a modified form of INOR-8 has the following composition: nickel base, molybdenum 15–18%, chromium 6–8%, iron 5%, manganese 1%, silicon 1%, phosphorus 0.015%, sulfur 0.02%, boron 0.01%, niobium 0.2%, hafnium 0.2%.

but the equilibrium pressure of SiF_4 is in general quite moderate.[232] When a small concentration of O^{2-} is compatible with the experimental aim, silica provides an inexpensive container which is very helpful in performing visual tests. Another electrical insulator material which is more corrosion-resistant is boron nitride,[233] but it is highly sensitive to moisture and has to be kept under a very dry atmosphere before use.

4.3.4. Aluminum Preparation

The great economic importance of the electrolytic preparation of aluminum has led to a great deal of laboratory research into cryolitic melts. At this small scale the most popular materials for melt containers are high-density graphites or nonporous carbons such as vitreous carbon. In the absence of dissolved metals (aluminum, alkali metals), carbonaceous materials are not wetted by liquid cryolite, nor are they destroyed by solidification of the electrolyte and thus they can serve for several experiments. Of course they require an oxygen-free atmosphere and it should be borne in mind that, at the usual working temperatures (about 1000°C), carbon dioxide reacts quantitatively with the carbon to give carbon monoxide.

When liquid aluminum or dissolved alkali metals are present in the bath, intercalation compounds or carbides appear on the surface of the carbon and great changes are observed in the interfacial tension[217] making the crucible more fragile. The same fragility occurs when the carbon or graphite crucibles serve as cathodes,[234,235] but in this case it is observed that metallic sodium is produced and condenses on the cold part of the cell.

Platinum has been successfully used for containers for cryolite baths,[236] its main advantage being its suitability when an oxygen-containing atmosphere is used. With metallic aluminum (produced, for example, during the electrolysis) alloy is formed.

Most of the insulator materials such as refractory oxides dissolve in molten cryolite. The only material that can be used is boron nitride, which does not react with liquid aluminum. However, as pointed out by Thonstad,[237] a rate of corrosion of 0.23 mg/cm^2 hr is observed for pyrolitic BN and in the presence of aluminum an additional attack occurs with the formation of aluminum nitride. When the anode gas comes into contact with boron nitride a reaction of the type

$$BN + 3/2\ CO_2 \rightarrow 1/2\ B_2O_3 + 1/2\ N_2 + 3/2\ CO$$

occurs;[238] thus the use of boron nitride for experiments requiring a very pure electrolyte is questionable. Because of the industrial interest in the mixture cryolite–alumina, many experiments have been carried out in these melts. With alumina-saturated baths (about 12% wt. at 1000°C) it is recommended to use a high density alumina crucible, which provides a nonpolluting and inexpensive container.[239]

4.3.5. Alkali Halides (Except Fluorides)

When the temperature is lower than 500°C Pyrex glass is the most popular material employed for the container of the melt; for higher temperatures, silica is suitable. These materials are very sensitive to the presence of traces of oxide ions which must be carefully removed. This can be achieved by repeated treatments with the corresponding hydrogen halides or halogens. When an atmosphere containing these active gases is used during the experiment no corrosion is observed. To remove the surface impurities all portions of the glassware coming into contact with the molten salt must be boiled in $HClO_4$, washed with deionized water and dried for several hours at 150°C.[240] However, even with dry and oxygen-free inert atmospheres (argon, helium, nitrogen), the glass surface becomes dull. At 500°C with Pyrex a more severe attack occurs which causes etching after a few hours. The ions of the melt, especially the lithium ions if present, slowly exchange with those of the glass (sodium) and an increasing concentration of oxide ions is observed in the fused liquid.[241] In the same way, a penetration of the alkali ions is observed in silica and, with this material, it is difficult to exceed a temperature of 900°C for more than a few hours. More resistant materials can be used such as alumina, zirconia, beryllium oxide, or metals such as platinum. Fused alkali halides are often used as solvents and, in this case, the aggressiveness of the solute must be taken into account, and when oxides, hydroxides, sulfides are dissolved it is better to use carbon crucibles. Nonporous graphite[242] now exits as well as vitreous carbon[243] which may be employed as melt containers. It is evident that a dry and oxygen-free atmosphere is required; above 500°C carbon dioxide may also react with these materials. Moreover, as with the Pyrex, the presence of dissolved alkali metals should be avoided, as the formation of alkali carbides is observed, and the crucible becomes very brittle. For the same reason, a cathodic polarization of the crucible is very dangerous. However, the carbonaceous materials may be used as anodes without damage.[244]

Experiments have been carried out with LiCl–$CaCl_2$ or LiCl–LiF melts in iron or mild steel crucibles at 700°C to study the behavior of dissolved calcium carbide or lithium nitride.[245] These materials are suitable also for the salts containing dissolved metal as in the preparation of the alkali metals.[123] The corrosion studies on various types of container materials in molten salt mixtures (NaCl–KCl, LiCl, $NaNO_3$, KNO_3, Na_2SO_4) have been carried out by Venkatasetty and Saathoff[247]—it has been shown that the presence of alkali nitrates considerably reduces the corrosion rate of iron or mild steel, an effect which is attributed to the passivating nature of the nitrate ions. For the electrowinning of zinc from $ZnCl_2$ the KCl–NaCl–$ZnCl_2$ electrolyte is contained in a crucible made of Sioxyn* which shows no sign of attack over a six-week period of electrolysis at 500°C.[248]

* Norton Trade Mark : silicon oxynitride.

4.3.6. Alkali Nitrates

Pyrex glass has been almost universally employed as the container material for these melts. But it is sensitive to the presence of oxide ions and care must be taken to ensure that the atmosphere is free from water vapor. Above the melting point nitrates dissociate according to the reaction: $2\ MNO_3 \rightleftharpoons 2\ MNO_2 + O_2$, which is reversible. At high temperatures, peroxides and oxides are formed. Lithium nitrate attacks glass containers above 300°C (Li^+ and Na^+ ions exchange) but the use of nitrous vapors considerably reduces this corrosion.[249] Below 400°C alumina or zirconia are not corroded by oxide ions, but when it is important to avoid the possibility of oxide ions entering the solution, platinum crucibles are recommended.[250,251] This metal is not suitable, however, when oxide ions are present since these ions are in equilibrium with peroxide ions ($O^{2-} + NO_2^- + O_2^{2-}$) which are known to attack platinum vigorously.[252,253] For experiments at low temperatures (below 280°C) with eutectic mixtures, Teflon is a suitable material for the salt container.[254]

4.3.7. Carbonates

Gold is the only metal completely immune to attack by the molten alkali carbonates.[255] Electrochemical tests show that the only change with time is grain growth.[256] No evidence of a gold oxide reduction peak is detected; this behavior is in contrast with that of platinum which shows a rapid attack with the formation of a greenish surface layer which, when Li_2CO_3 is present in the melt, is presumably lithium platinate. Gold–palladium alloys have also been recommended.

Porcelain and refractory oxides such as corundum, alumina, and magnesia have been used but they bring serious corrosion problems which depend mainly on the experimental conditions; indeed, in an inert atmosphere at the usual operating conditions molten carbonates decompose to give CO_2 and oxides.[257] Then it is preferable to operate in a CO_2 atmosphere. For this reason, many laboratory experiments are carried out in a hermetically sealed cylindrical silica tube containing the crucible.

When economic considerations are important, it is not possible to seriously envisage the use of gold or gold alloy. A selection of more common container materials for alkali metal carbonates has been recently reviewed by Grantham and Ferry.[258] At 500°C austenitic stainless steel is satisfactory; at 600°C nickel-based alloys are required, while at 700°C high-chromium alloys have to be used. No metals or alloys have been found that provide satisfactory containers above 800°C. High-density alumina is corrosion-resistant at these temperatures, but corrosion is highly dependent on impurities such as fluorides or hydroxides. The cermet LTl (Union Carbide) has shown a remarkable corrosion resistance up to 1000°C. AlSl 316 stainless steel is presently used as the material of construction for the molten carbonate fuel cells,[259] but catastrophic corrosion is observed in the more oxidizing environment (CO_2 outlet region). Nickel and copper-base coatings are being considered.

4.3.8. Sulfates

At low temperatures with the ternary mixture Li_2SO_4–Na_2SO_4–K_2SO_4 Pyrex crucibles have been used,[260] but severe attack is rapidly observed and silica is preferable.[208] However, these materials are not convenient when oxide ions are present. Melt containers of platinum or of recrystallized alumina (silica free) are not perceptibly attacked by fused sulfates containing added oxides; they support also an oxygen-containing atmosphere in which these melts are stable.

4.3.9. Hydroxides

Mixtures of alkali metal hydroxides exhibit low melting points (melting point of the eutectic mixture NaOH–KOH : 170°C). Below 280°C Teflon crucibles have been successfully used;[261,262] however a chemical attack occurs when the melt contains dissolved alkali metals. The most popular material is vitreous carbon which can be used at high temperatures but, as already pointed out, this material is not stable in the presence of dissolved alkali metals. Ceramic materials and alumina withstand these melts in the absence of oxide ions or moisture.

Studies of the reactions between metals and molten hydroxides show that up to 500°C nickel does not react if oxygen is excluded from the atmosphere;[263] small amounts of oxygen induce the formation of a black coating of NiO_2 which is insoluble (less that 0.5 mg of Ni ions are determined in NaOH at 500°C).[264] However, this coating dissolves in basic melts (presence of oxide ions).

Electrochemical studies indicate that platinum behaves analogously to nickel but is less reducing, and thus less resistant to oxidizing basic melts. Although the use of a gold crucible has been reported,[265] it seems that this metal is rapidly attacked, mainly when oxide ions are present. Silver is more difficult to oxidize and thus more resistant; however, when alkali metals are dissolved in the bath the crucible is rapidly destroyed (formation of silver-alkali metal alloys).

5. Temperature Measurement

5.1. Introduction

Ionic compounds exhibit in general a very good thermal stability, and they provide a suitable basis for research in the high-temperature field. The temperature range in which the simple fused salts, such as alkali halides, are stable is very large; from about 320°C for the binary or ternary eutectic mixtures, up to the boiling points ($\theta_E = 1413°C$ for NaCl). They are of value to determine the influence of temperature on the physico-chemical properties or on the reaction rates. The accurate measurement of these properties requires the accurate measurement of the temperature; for example, a fluctuation of ±1°C corresponds to a relative error of about 0.3% on the value of the self-diffusion coefficients in alkali halides at about 400°C.[266] This accurate determination is the more important now as those ionic liquids constitute a fundamental basis for the development of the theories of transport in the liquid state.[267,268]

5.2. Definition

The basic definition of the temperature is drawn from the laws of thermodynamics and can be illustrated by the aid of the Carnot heat engine : the ratio of two temperatures of heat reservoirs is defined as the heats absorbed from the two reservoirs. The temperature scale based on an ideal gas is the same as the thermodynamic temperature. Many experimental arrangements have been given, but they are not of practical use and they serve only to measure the temperatures of standard points. The International Practical Temperature Scale is based on a number of fixed and reproducible points.[269]

Primary fixed temperature points (at 1 atm) in °C.

Water triple point.	0.01
Liquid water and its vapor.	100.
Solid and liquid zinc.	419.58
Solid and liquid silver.	961.93
Solid and liquid gold.	1064.43

Other thermometry fixed points.

Ice melting in air saturated vapor (ice point).	0.00
Freezing point of tin.	231.97
Freezing point of lead.	327.50
Boiling point of sulfur.	444.67
Freezing point of antimony	630.74
Freezing point of palladium.	1554.

5.3. Platinum Resistance

Between the above fixed points and over the range −183°C and +630.5°C the scale is defined by the platinum resistance thermometer. The resistance R_T is related to the temperature by the quadratic equation:

$$R_T/R_0 = 1 + A\theta + B\theta^2$$

where R_0 (resistance at the ice point), A, and B are constants determined by calibration at the water-triple point, the steam point (100°C), and the freezing point of zinc. Many commercially available apparatuses have a platinum resistance of 100Ω at 0°C. A three-wire connection is generally used to eliminate the influence of the resistance circuit.

The accuracy of the measurement is ±0.01°C. These thermometers can be used at higher temperatures up to 1000°C. The changes that may take place in a platinum resistor are of two basic types: changes in the dimension of the wire and changes in the purity. Prolonged heating at high temperatures produces recrystallization and crystal growth. These changes have only a small effect on the resistance; however, they may weaken the wire mechanically making it more susceptible to work hardening. In commercial thermometers, the platinum resistor is mounted on a support made of silica and has to be used in an oxidizing or inert atmosphere. Reducing agents such as carbon or hydrogen may react giving platinum silicides. In general, it is of value to calibrate the thermometer after a long period of time especially when it has been used at temperatures higher than 400°C. This calibration can be performed by comparing the readings from the thermometer with those of a similar thermometer the calibrations of which are accurately known; special furnaces with a uniform temperature zone have been designed for this purpose.[270]

5.4. Thermocouple

The thermoelectric effect has been employed for over 140 years as a practical tool for the measurement of temperature. When a circuit is formed by two dissimilar metallic conductors A and B, and one of the junctions of A and B is at a temperature T_1 while the other is at a higher temperature T_2, a current will flow in the circuit. It has been established that there is a definite relationship between the difference of temperatures of the junctions of a thermoelectric circuit and the resultant emf in that circuit. The introduction of a third metal in the measuring device has no effect so long as its junctions with A and B are both at the lower temperature.

It has been found that the couple platinum alloy of platinum with 10% rhodium gives a thermoelectric power curve which is nearly linear from 400 to 1100°C. It is, from the scientific standpoint, the most important of the thermocouples now in use. It is used for defining the International Temperature from 630.74°C to 1064.43°C (gold point). The interpolation equation giving the emf E is $E = a + b\theta + c\theta^2$ where the constants are fixed by calibrations at 630.74°C determined by a platinum resistance thermometer and at the freezing points of silver and gold.

When it is desirable to measure temperature outside the above range without changing the measurement method, the platinum thermocouple is a useful tool. It may be used in the range from 0°C up to 1500°C and, with special precautions, for short periods even up to 1650°C. At that temperature the high purity platinum forms large grains; this grain growth results in a more fragile wire so that mechanical failure can readily occur and, moreover, both vacancy trapping and production of rhodium oxide in the Pt–10% Rh leg can cause significant reduction in the thermoelectric power.[271]

To obtain reproducible results, it is important to avoid contamination which may result from the environment, the furnace, and the materials used in the insulators.[272]

Silicon hastens failure by forming a eutectic alloy which has a melting point of 800°C and enters grain boundaries. Prior to fracture, the beginning of penetration is indicated by erratic swings of the emf. Thus, for temperatures above 1000°C, it is recommended to avoid insulating and protective tubes made of quartz or of materials having a high silica content (such as synthetic mullite). Sintered or vitreous alumina may be useful, but this material has the least resistance to thermal shocks. Magnesia is the usual refractory for swaged thermocouple assemblies, as it packs better than alumina. When an external protection tube made of platinum is used certain precautions must be followed. It is better to use an inert atmosphere. In an oxidizing atmosphere (air) heating above 1000°C produces weight losses which are attributed to the formation of volatile oxides. Hydrogen has no deleterious effect, but this gas may reduce some of the impurity oxides in the environment which then attack the metal. There is a reaction between platinum and H_2S above 400°C and a blue film appears on the surface. No attack is reported by CO, CO_2, SO_2, SO_3. Platinum is attacked by sulfur vapor and by metals such as zinc or lead. Direct contacts between platinum and alkali or alkaline earth hydroxides should be avoided.

5.5. *Measurement of the Temperature in an Electro-Chemical Cell*

The measurement of temperature is significant only when two conditions are fulfilled: the temperature must be stable and the temperature representative of the studied phenomenon has to be the same as that of the sensing element (topographic homogeneity). The aim of the design of the heating arrangement is to satisfy these conditions, by introducing for example a metal liner in the hot zone of the furnace. The temperature distribution depends also on the heated body itself; ionic liquids constitute favorable media to obtain uniform temperatures as their thermal conductivities and heat capacities are quite large and the convection process occurs freely. However, in electrochemical cells the studied phenomenon depends on the temperature of the electrode surface or at least of a very thin layer near the electrode surface. Great care should be taken to ensure that the electrode temperature is correctly monitored. When no special precautions are taken a difference of several degrees between the bath and the electrode can appear: this has been determined by the use of hollow electrodes with a thermocouple inside, or by the determination of the freezing point of the electrolyte at the electrode surface. A compromise must be found between the electrical resistance of the electrode support (which has to be as small as possible to reduce the ohmic drop) and its thermal conductance. Additional heating of the electrode support has been used but in general the most popular arrangement is to place the bath level just below the maximum temperature in the hot zone of the furnace to compensate the heat loss by thermal conductivity towards the upper part of the electrode support. Shielding of the current supply in the hot zone of the furnace has to be avoided.

The heat dissipated by the Joule effect must be taken into account. As pointed out by Mergault,[273] this dissipation creates a large distortion of the temperature distribution near the electrode surface; when high current densities are used this heating produces an overheating of the electrolyte and the so-called "electrode effect" occurs.[274]

5.5.1. Other Thermometers

The two thermometers described previously (platinum resistance and platinum alloy thermocouple) give very accurate and stable measurements; however, they require high performance electronic devices (amplifier, impedance bridge, precision potentiometer). More accessible apparatuses can be used when a high degree of precision is not required.

5.5.2. Thermocouples

The thermoelement combinations most commonly used are Chromel-Alumel and iron-Constantan. Constantan is an alloy of copper and nickel, $Cu_{57}\ Ni_{43}$, with the addition of small percentages of Mn and Fe. The thermoelectric power is about 63 μV/°C at 800°C. This couple is serviceable in oxidizing atmospheres at temperatures up to about 760°C where it has a life of about 1000 hours. It may also be used in reducing atmospheres up to about 900°C.

Chromel is an alloy having the composition $Ni_{90}\ Cr_{10}$, and the Alumel, developed with the special purpose of serving as the negative element of the couple, has the composition $Ni_{94}\ Mn_3\ Al_2\ Si_1$. The thermoelectric power of the couple is fairly constant at about 40 μV/°C over the range from 250°C to 1000°C, and its emf is more than four times greater than that of the Pt/Pt-10% Rh thermocouple. The oxidation-resistant characteristics of the couple are better than those of other common metals couples and permit its continuous use at temperatures up to 1000°C in a clean oxidizing atmosphere. The emf deterioration of the Chromel-Alumel thermocouple in reducing gas can be prevented by a ceramic boron nitride coating.[275]

5.5.3. Reference Temperatures

It is a common practice to reference both legs of the thermocouple to copper leads at the ice point; then the copper leads may be directly connected to the readout instrument (See Figure 12). As the emf generated is dependent on the temperature difference it is important to obtain a stable and well-defined reference temperature. As ice baths are not always practical to maintain, alternative methods are often employed.[276]

Self-compensating electrical leads are used—they incorporate a temperature-sensitive resistance R_s which has to be integrated with the cold junction T_H (Figure 13).

The compensation accuracy of the commercially available cold junction compensator is about ± 0.5°C from 15°C to 30°C.

A new ice-point reference chamber has been developed which completely eliminates the need for the frequent attention required by the common ice-bath; the outer walls of the reference well are cooled by thermoelectric cooling elements to cause freezing of very pure water in the cell. The accuracy is from 0° to $+0.1$°C in the ambient temperature range from 2° to 32°C.

Of course when the cold junction is at a stable temperature (ice-point bath) or when a self-powered thermocouple reference is used, it is important to make sure that the readout instrumentation is uncompensated.

5.5.4. Thermistor

Some partially metallic elements (carbon, silicon, germanium) and many oxides or oxide mixtures are semiconductors. Their resistance decreases with increasing temperature and at a much faster rate in contrast to that of metals. Today, thermistors are usually made of a polycrystalline mixture of several metal oxides such as manganese, iron, cobalt, nickel, copper, magnesium, and titanium.

Thermistors have very high temperature sensitivities.[277,278] Accuracy is limited in most applications only by the readout device. The following equation has been proposed to represent the thermistor behavior[276]:

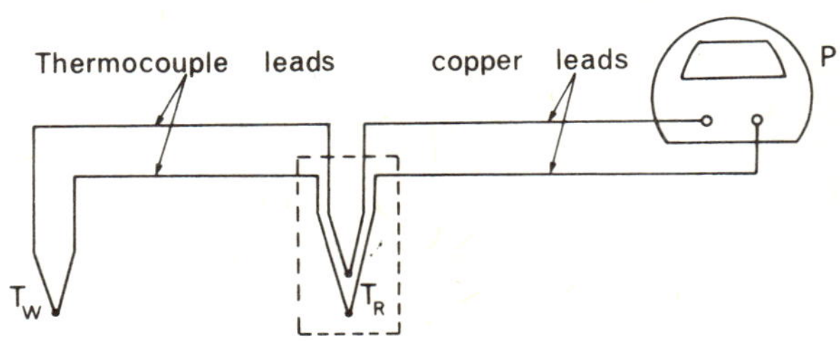

FIGURE 12. Typical thermocouple circuit; T_W: measuring junction; T_R: reference junction; *P:* potentiometer.

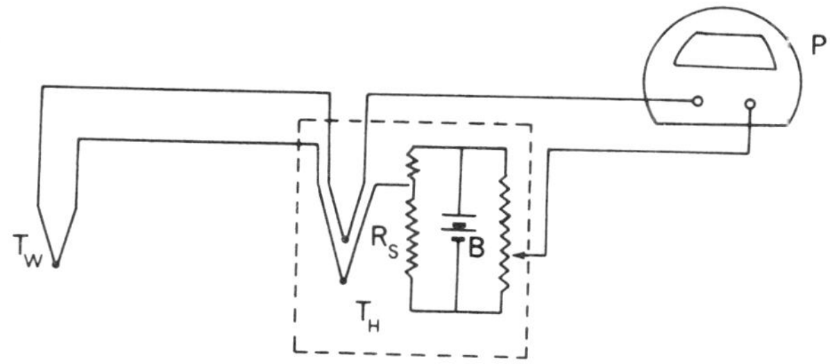

FIGURE 13. Thermocouple circuit with a cold junction compensator; R_S: sensitive resistance element; B: battery; T_H: chamber at a homogeneous temperature.

$$1/T = A + B \operatorname{Log} R + C(\operatorname{Log} R)^3$$

Where T = °Kelvin, R = resistance, and A, B, C = fitting constants. Interpolation data generated by this equation are accurate to ±0.01°C.

Recently, a single crystalline intrinsic semiconductor has been used as a resistance thermometer material. For the temperature range 30–180°C the best semiconductor is high-purity germanium.[279] The advantage of Ge thermistors lies in their high precision at relatively low cost and good miniaturization.

Since resistance values of thermistors vary up to the megohm range, the effect of the exterior wire resistance is lessened. Thus there is no need for external compensation or multiple conductor cables.

The main disadvantage of thermistors, however, is their instability in the high temperature range which limits their use at temperatures of about 290°C;[280] but with low melting mixtures, hydrated fused salts, acidic salts such as 2HF–KF, and organic fused salts, they may give a very suitable temperature measurement.

5.5.5. Mercury Thermometers

The mercury thermometer is an extremely simple instrument, cheaper and requiring less auxiliary apparatus than other temperature measuring devices, but at high temperatures some irreversible changes can appear. For example, the zero of a borosilicate-glass thermometer rose at the rate of 0.04°C/100 hours when heated at 460°C.[281] Higher temperatures can be measured using a quartz thermometer; most of these thermometers range from 100 to 600°C. A very good stability is observed, but in general calibration curves are necessary. This is mainly due to the variations in the cross section of the bore, but given a good supply of capillary, an accuracy of a few hundredths of a degree at 600°C seems usually to be obtained. In using quartz thermometers at high temperatures, care must be taken to avoid chemical contamination which may lead to devitrification (volatile alkali for example).

5.5.6. High Temperature Measurements

There is now an increasing interest in electrochemistry at high temperatures which opens the way for electrolytic preparation in molten silicates[282,283] or oxides.[284] At this level, the temperature measurements become crucial. Above the gold-point the temperature is defined by Planck's law for radiation from a black body. The most popular instrument used to realize this measurement is the disappearing filament optical pyrometer but, in general, it does not allow an easy measurement in electrochemical cells where thermocouples are often the only practical means of temperature measurement. As pointed out above, the Pt/Pt10%Rh couple may be used continuously up to 1500°C (fluctuation of the emf of 45μV is reported at that temperature after 20 days). The couple Rh/Pt–20%Rh seems to have a better stability. For continuous use at temperatures above 1500°C, the tungsten-rhenium thermocouples are recommended.[285] These alloys have a poor oxidation resistance and should be used in vacuum, hydrogen, or inert atmospheres.

At these temperatures, beryllia or thoria are the best insulators. Of all refractories, thoria is attacked the least by molten metals; on the negative side, it has poor thermal shock resistance, radioactivity, and high cost.

Increased sensitivity can be obtained by connecting several thermocouples in series to form a thermopile; for measurement above 800°C such a pile has been developed using the couple 94 Pt–6Rh/70Pt–30 Rh.[286] Its advantage is its extremely low sensitivity to the cold-junction temperature changes. If properly fabricated (alumina multibore insulator tube and platinum shielding), it has a

TABLE IV. Comparison of Temperature Transducers

	Thermocouples	Resistance	Thermistors
Sensitivity	10 to 50μV/°C	0.2 to 10Ω/°C	100 to 1000Ω/°C
Stability	1° to 2° drift per year	Less than 0.1% drift in 5 years	0.2° to 3°C drift per year
Temperature range	−240°C to 1300°C[a]	−130°C to 850°C	−100°C to 290°C
Interchangeability	2 to 8°C	0.05°C to 0.1°C	0.2° to 2°C
Features	Require the preservation of the cold junction at a standard reference temperature, greatest economy	Greatest accuracy over wide span	Greatest sensitivity

[a] Extended ranges to over 2700°C with tungsten–rhenium couple.

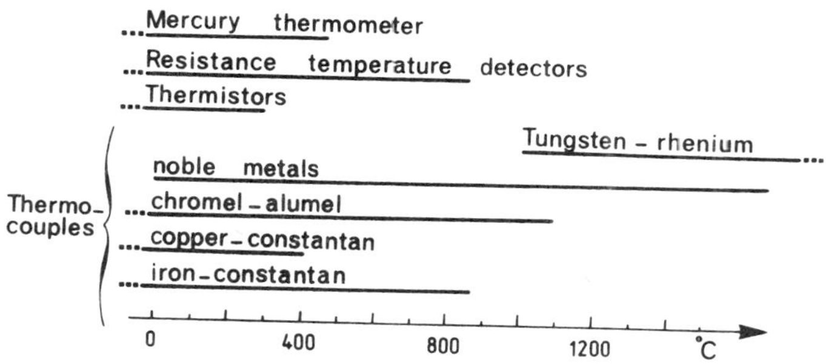

FIGURE 14. Useful range of some thermometers.

long life at temperatures up to 1600°C. A six-junction thermopile made of this material produces higher emf at high temperatures than the sensitive Chromel-Alumel thermocouple which simplifies the apparatus needed to obtain temperature measurements of a given precision.

Summary

The properties and special features of thermocouples, resistance thermometers, and thermistors are compared and contrasted in Table IV and Figure 14, in which the useful temperature ranges are highlighted.

6. *Temperature Control*

6.1. *Introduction*

In general, the electrochemical experiments are carried out at a controlled temperature and it is thus necessary to use a technique which will provide the required degree of control. This requirement varies according to the experimental purpose and the nature of the measured quantities. For example, some data such as transport numbers are not much affected by temperature changes, but the absolute measurement of the ionic mobility, of the diffusion coefficient, or of the conductivity requires a precise temperature; a variation of ± 1°C induces a change of about $\pm 0.3\%$ in the quantities. The emf measurements are not very dependent on the temperature: about ± 0.5 mv for a variation of 1°C, but these measurements are important for the determination of the thermodynamic properties of many systems and thus require accurate temperature control.

6.2. Manual Control

The simplest technique to obtain a stable temperature is to use a constant voltage supply; the furnace will reach its steady-state temperature T_M corresponding to this constant power supply. To illustrate the behavior of such a system, let us consider an isothermal body at the temperature T isolated from the external medium by a coating of thickness e (Figure 15). The rate of heating dT/dt at a constant power input P is given by the heat flow equation:

$$P = \frac{dH}{dt} = C_P \frac{dT}{dt} + \frac{D\, s\, \rho\, S}{e}(T - T_a) \qquad (1)$$

where H is the heat input, C_p the heat capacity of the body, D the heat diffusivity through the coating; s and ρ are respectively the specific heat and the density of the coating; S and e represent the mean surface through which the heat flows and the thickness of the coating; T_a is the temperature in the external wall. Assuming that T_a = the room temperature and C_p and D are supposed constant, the temperature variation of the heated body is represented by:

$$T = T_M - \frac{P\, e}{D\, s\, \rho\, S} \exp(-\, t/\tau) \qquad (2)$$

with $\tau = \dfrac{eC_P}{D\, s\, \rho\, S}$ and $T_M = T_a + \dfrac{P\, e}{D\, s\, \rho\, S} = T_a + \dfrac{P}{C_p}\tau$. For a large furnace with very good insulation (small ρ, s, and D) the time constant τ may reach some hours; and according to equation (2) we have:

$$\frac{T_M - T}{T_M - T_a} = \exp\left(-\frac{t}{\tau}\right) \qquad (3)$$

Thus, to obtain a temperature T near the equilibrium temperature T_M a waiting time longer than τ is necessary. For example, to obtain a difference $T_M - T$ which is less than 1% of $T_M - T_a$ it is necessary to wait 4.6 τ.

Small electric furnaces can be built to run directly from the mains supply

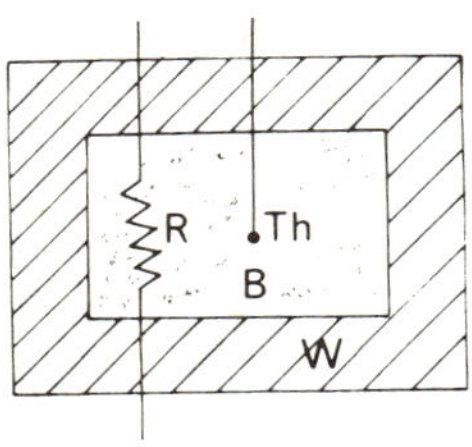

FIGURE 15. Furnace control, isothermal representation; R: heating element; Th: therometer; B: isothermal body; W: surrounding wall.

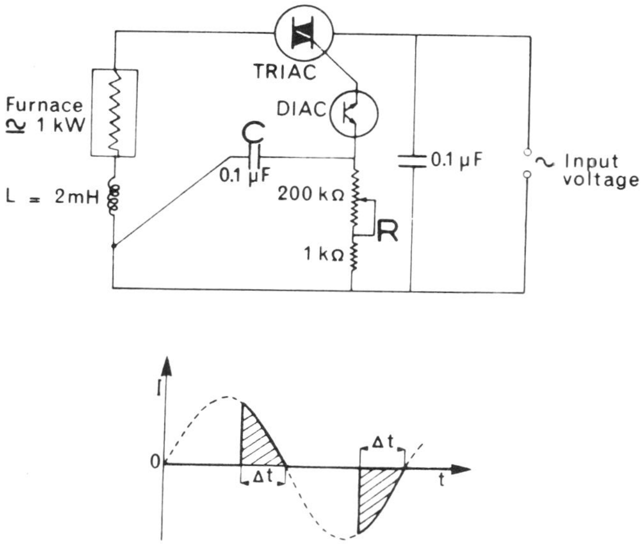

FIGURE 16. Diagram of power adjustment using a discontinuous fraction of time control. Representative curve of the feeding current.

and hence have a fixed equilibrium temperature; however, a manual control of voltage input is adequate for many applications. Resistance heaters are regulated by means of rheostats, time interruption switches, and voltage transformers. The rheostats which are connected in series with the heater have the advantage of being relatively cheap, but they introduce a large power loss and the range of operating temperatures is limited. A more efficient method is to use an energy regulator which periodically switches the furnace on and off. By providing a variable ratio of time on to time off, it varies the heating rate and the equilibrium temperature; to avoid temperature variations of the heated body the period of the "on-off" system must be much shorter than the time constant τ of the furnace. A current chopper based on the frequency ν of alternating current can be provided by a silicon bidirectional triode thyristor or triac. This device is designed to switch from "off" to "on" for either polarity of applied voltage with positive or negative gate triggering voltage. A typical control circuit and the waveform of the current load from a sinusoidal mains voltage are given in Figure 16. The current flows through the furnace during a time Δt, the quantity $\Theta = 4\pi\nu\Delta t$ is called the conduction angle and varies according to the position of the potentiometer of the phase shifter. The high speed switching of the current (a few microseconds) requires presence of a low pass filter (R and C, Fig. 16).

An efficient current adjustment can be provided by Field Effect Transistors with a Vertical Metal Oxide Semiconductor structure, VMOS[287,288] or HEX-

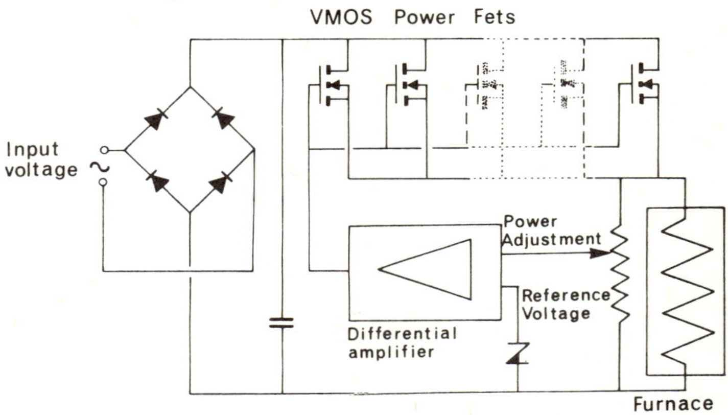

FIGURE 17. Diagram of voltage adjustment using power transistors.

FET.* They are used with direct current, but power rectifiers are now easily available. This structure gives several important advantages over bipolar devices; it has a very high input impedance and thus MOS will directly interface with high impedance drivers. Moreover, it has no secondary breakdown voltage and, since the temperature coefficient is positive, it draws less current as the device heats up. Then, if for some reason the current density tends to increase in the external circuit, the temperature rises and the current is automatically limited, so hot spot or current crowding cannot develop. The current is automatically shared between parallel devices (Figure 17) so no ballasting resistors are needed. However, with this power transistor a heat loss occurs which requires the use of a suitable heatsink.

It is also possible to control the power input by periodically switching out

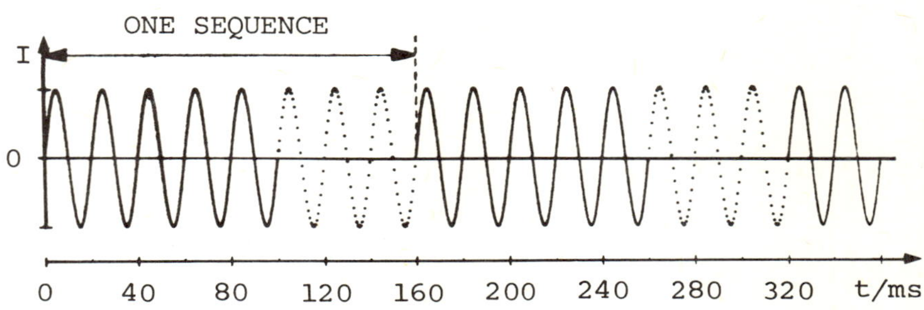

FIGURE 18. Waveforms of heating current pulsation. Five cycles of current flow followed by three cycles of current absence (frequency: 50Hz).

* International Rectifier—Trade Mark.

a certain number of cycles of current.[288] Figure 18 shows a graph of current versus time; switching at zero voltage makes it possible to switch large currents (2500 A and over) without generating radio interference or introducing harmonics into the grid. By simultaneous use of several units, it is possible to obtain a good power factor. The low response-time regulation mode (basic switching at 20 ms) eliminates flicker and thermal ripple. Moreover, the power dissipation is very small.

Variable voltage auto-transformers provide a convenient means of control but they are expensive when compared with the above-mentioned systems and power losses are negligible. A multi-tap transformer in conjunction with a variable resistance can also be very useful. Some precautions must be taken with transformers; for example, two transformers must never be connected in parallel, i.e., to the same resistor, as high amperage circulating current can develop.

Manual control using a fixed power input is convenient when some temperature fluctuation can be tolerated (preliminary experiments, chemical treatments, purification, etc.). Indeed, the normal variation in power line voltage (220 ± 10 volts) may be sufficient to cause large temperature fluctuations and also the heat loss of the furnace can vary (change of the room temperature, draughts, etc.).

6.3. Automatic Control

An automatic temperature control system consists of a temperature sensing-device to compare the actual temperature with the desired or set temperature and a means of altering heat input to correct deviations of actual temperature from the set point. The temperature level is in general recorded by an electrical unit; the deviation of this reading from the set point is used to change the thermal input and attain the required temperature. This is done in two ways.

6.3.1. A Two Position Control

In this type of control, the power is switched between two fixed values (high–low control) P_{M} and P_{m}. The power input will be a step-function of time. When the control is in steady operation, the furnace temperature oscillates between two limits T_1 and T_2. To understand the behavior of this control, we refer to the previous scheme (Figure 15). According to Equation 2, the temperature variations are represented by the equations:

$$\text{Rising curve: } T = T_{\mathrm{M}} + (T_1 - T_{\mathrm{M}}) \exp\left(-\frac{t}{\tau}\right) \tag{4}$$

$$\text{Descending curve: } T = T_m + (T_2 - T_{\mathrm{m}}) \exp\left(-\frac{t}{\tau}\right)$$

where T_M and T_m represent the limit temperature ($t \to \infty$) corresponding to the power inputs P_M and P_m (Figure 19).

The durations of the ascending and descending parts are, respectively:

$$t_{12} = \tau \ln \frac{T_M - T_1}{T_M - T_2} \text{ and } t_{21} = \tau \ln \frac{T_2 - T_m}{T_1 - T_m} \tag{5}$$

It is convenient to choose the power inputs so that the mean temperature $(T_M + T_m)/2$ is equal to the set temperature T_θ—then $T_1 + T_2 = T_M + T_m = 2T_\theta$; in this case $t_{12} = t_{21}$. When the temperature difference $\delta T = T_2 - T_\theta = T_\theta - T_1$ (which can correspond to the sensitivity of the sensing element), is small with respect to the difference $2\Delta T = T_M - T_m$, we have:

$$t_{12} = t_{21} = 2\tau \frac{\delta T}{\Delta T} \tag{6}$$

For a given ΔT and a given furnace (which determines τ) it is interesting to reduce the duration of the oscillations which decreases amplitude δT. This may easily be achieved by placing the sensing element near the resistor (Figure 20), the heat capacity of the region around the element being smaller than that of the heated body; the time constant τ which is now used to regulate the furnace is greatly reduced. Sometimes an auxiliary heater around the sensing element is used; the current of this heater is switched on and off by the thermostat in step with the main heating element.[289]

The main cause of temperature oscillation is heat propagation. Indeed, owing to the separation of the resistor and the temperature detector, the change in heat production is not sensed until a finite time has passed. Assuming that the heat flow is in the x direction only, the continuity equation is written:

$$D \frac{d^2T}{dx^2} - \frac{dT}{dt} = 0 \tag{7}$$

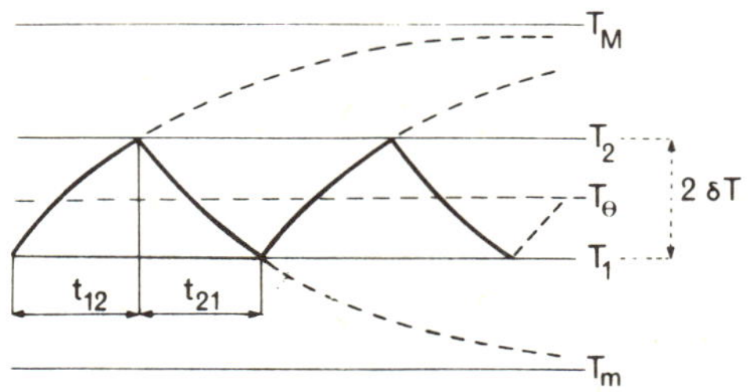

FIGURE 19. Temperature oscillation in two-position control.

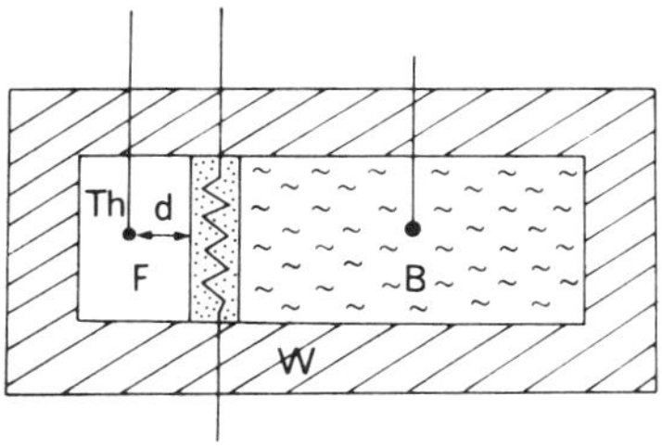

FIGURE 20. Furnace control with the sensing element Th, in a region F, near the heating element R (or heated by an additional resistor in derivation with the main resistor); B: heated body.

It will be assumed that the temperature of the heating wall (near the resistor) is a periodic function:

$$T(x = o,t) = \Theta \sin \frac{2\pi t}{\tau_c} + T_\theta$$

Then the solution of Equation 7 for a semi-infinite homogeneous medium[290] is:

$$T(x,t) = \Theta \exp \left(\frac{-x}{\sqrt{D\tau_c/\pi}} \right) \sin \frac{2\pi}{\tau_c} \left(t - \frac{x}{2} \sqrt{\frac{\tau_c}{\pi D}} \right) + T_\theta \tag{8}$$

Figure 21 gives the temperature profile at the maximum temperature of the wall $t_0 = \tau_c/4$ and at an instant t_1 later. The curve displaces on the right and deforms; the maxima and minima of temperature displace on the two exponentials $\pm\Theta \exp(-x\sqrt{\pi/D\tau_c})$. The change of the power input $P_M \rightarrow P_m$ will occur every time the temperature on the sensing element passes upwards through the set temperature T_θ.

Thus the oscillations of the temperature will depend on the distance d between the heating resistance and the sensing element; as the oscillations on the sensing element will be one quarter period behind those on the resistor. Equation 8 gives:

$$d = \sqrt{\pi D\tau_c/4} \tag{9}$$

At this point, the temperature oscillations are $\pm\Theta\exp(-\pi/2)$ around T_θ. It is shown that this distance d should be as short as possible to reduce the period τ_c which is about proportional to the amplitude Θ. Indeed, as pointed out by Motzfeldt,[291] it can be assumed that the temperature on the resistor follows Equation 4 when the power input is constant, i.e., during the interval $\tau_c/2$ which corresponds to the time t_{12}:

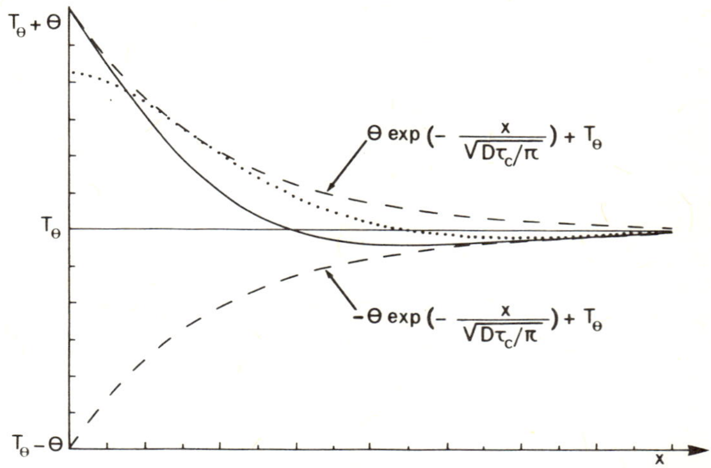

FIGURE 21. Temperature profile in a homogeneous semi-infinite medium against a heated wall with a sinusoidal variation of the wall temperature. The full line represents the temperature profile at the maximum temperature of the wall; the dotted curve shows the evolution of this profile an instant later. The temperature at one point oscillates between the two dashed curves.

$$\frac{\tau_c}{2} = 2\,\tau\frac{\delta T}{\Delta T}$$

Then

$$\Theta = \delta T = \frac{\tau_c}{4\tau}\,\Delta T \tag{10}$$

The time constant τ is determined by the nature of the furnace and this equation indicates that to minimize temperature oscillation the period τ_c has to be a small fraction of the furnace time constant; Motzfeldt indicates that an order of magnitude $10^{-3}\tau$ is desirable. Also the interval ΔT has to be as small as possible; thus the higher and lower power inputs have to be chosen as closely as possible above the limit of the fluctuations of the mains voltage and heat load.

Instead of regulating the entire power supply, it is often convenient to control only a small auxiliary heater which reduces the amplitude of temperature oscillation. Owing to the heterogeneous structure of the furnace, Equation 8 is only a rough approximation of the real system; a very large temperature gradient may exist within the furnace; a reservoir of high capacity and good thermal conductivity should be present. Molten salts are the best liquid for this purpose[292] and it is only necessary to ensure that adequate stirring is provided.

The change in power input of the furnace can easily be carried out by introducing an extra resistance which is shortcircuited when the temperature is too high. With a multi-tap auto transformer, a two-position voltage can be used to regulate the furnace and avoid the heat loss in the additional resistor.

Many commercial electronic switching devices are available. Dry or mercury contacts are used and a temperature regulator can be easily constructed from a conventional potentiometric recorder (Leeds and Northrup);[293] a second potentiometer is linked to the shaft of the main potentiometer, the input available to the furnace being monitored by a triac semiconductor or a power VMOS.

6.3.2. Proportional, Integral, and Derivative Control

To obtain a good constant temperature, it is of value to adjust the power input in proportion to the deviation from the net temperature. As shown previously, a change in the heat production will not be detected by the sensing element until a finite time has passed when, if the power input is too large (high proportional factor), at some moment strong oscillation may occur.

The linear regulation is maintained in a given temperature interval around the set point. For temperatures outside this proportional band limitation occurs and the control passes to the "on-off" mode. If the band width is widened, the change occurs more slowly (decreased gain), but for too narrow band widths the system may oscillate increasingly; these erratic oscillations can be avoided by a suitable adjustment of the automatic rate-circuit which acts as an anticipating device. An automatic rate-circuit[276] (derivative) is used to anticipate sudden line, load, or process upset and applies immediate correction. For example, for slow reacting systems (long term constant) this device induces a larger corrective output change than with proportional action only.

In all proportional controllers, the set point is in the center of the proportional band and corresponds to 50% input power. However, to maintain the controlled temperature, a different percentage of the input power is generally required and it is then necessary to include an automatic reset device (integral) which will move the band so that the required percentage of input power is at the set point.

The more common type of proportional control for electric furnaces employs a discontinuous fraction-of-time control such as the triac; the conduction angle of the triac and thus the averge power input are controlled by varying the phase displacement of the gate voltage (see above). A servo-motor working on a variable transformer has been used.[294]

A saturable core reactor can constitute a very good control element; a dc winding controls the impedance of an ac coil by varying the degree of saturation of the iron core. The initial cost of this element is not very different from the other instruments and its life expectancy is greater.

It is also possible to supply the heating element with a rectified current and to include in the circuit a ballast power transistor (HEXFET, VMOSFET). These elements can control very high intensities and they require very low pilot power; with drive currents as low as a fraction of μA it is possible to control several hundred amperes in the main circuit. As pointed out above, they are very robust (their temperature coefficient is positive) and they can be directly monitored by the logic circuit which follows the sensing element.

The tendency is now to use the sequential switching of the current at zero voltage (Figure 18) which gives the possibility of controlling large currents with a very good efficiency and without generating electrical interferences.

Another tendency developed around MOSFET and bipolar power transistors is the use of fast switching (200 kHz and above) power supplies[295] which allow the use of light low-cost transformers.

7. Furnaces

7.1. Introduction

Most of the experiments in fused salts need to be carried out in a temperature-controlled environment. This requirement is in general not very hard to meet, since reliable and accurately controlled thermostats and furnaces are now standard pieces of laboratory equipment. They can be purchased complete or built by the research worker, the latter solution, which is cheaper and probably more rapid, giving a better adaptation to the specific requirements of the electrochemical research.

In selecting a furnace for a particular application, it is important to be aware of its potentialities, cost, and probable life. Electric heating offers better possibilities for control than do other methods, and it is now universally adopted when a hot, uniform, and constant temperature is required. From the standpoint of laboratory research, the best type of furnace is an electrically heated tube. Large molten salt cells can be maintained at the working temperature by internally generated ohmic heat; studies of experimental self-heating cells have been carried out recently[23] and induction heating has been used in a few experiments.[97] Most of the research work in the fused salt studies is carried out in electrical resistance furnaces.

Important considerations in this type of work are the temperature control, the presence of a large enough zone of uniform temperature into which the cell can be introduced, an oxidizing, neutral, or reducing atmosphere, facilities for accurate measurement of the temperature, and easy access to the heated chamber (for the electrical leads, the tubings required for the gas bubbling, sampling of the electrolyte, etc.).

7.1.1. Electrical Resistance Furnaces

Resistance-heated furnaces are most commonly employed in laboratory practice and consist of a refractory chamber heated internally or externally by rods or windings and suitably insulated.

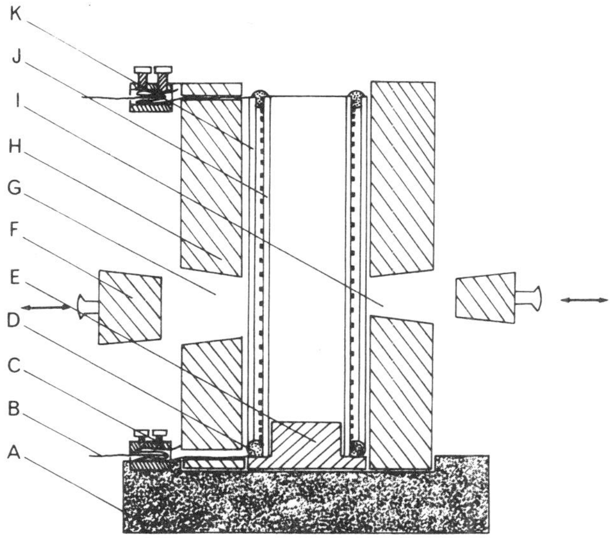

FIGURE 22. A general purpose laboratory furnace for a maximum operating temperature of 500 °C. A: Refractory slab; B: Lead wires; C: Screw terminal blocks; D: Asbestos and cement; E: Refractory block; F: Window shutting; G: Window for seeing through the cell; H: Silico-alumina felt; I: Aperture to illuminate the cell; J: Inner Pyrex tube with the nichrome resistor; K: Outer Pyrex tube to protect the resistor.

Details are given below for a furnace[296] which has been in general use in the laboratory of one of the authors (Figure 22).

The metallic heating element is a ribbon of Nickel-Chrome alloy (Metal Imphy, Ni–Cr 80–20; width: 1.6 mm; thickness: 0.2mm). It is wound on a Pyrex tube (I.D. : 80 mm; thickness: 2 mm; length: 38 cm). The usual procedure in winding the tube is first to secure the ribbon at one end with a wet asbestos string (diam.: 2mm) and then to form the ribbon spiral by slowly rotating the tube (mounted for example on a lathe). The space between two spirals is about 7.5 mm in the central part of the furnace. This distance is reduced progressively at the extremities of the tube to compensate for the heat loss which occurs at the open ends. On the upper part, the interval is decreased by 0.3 mm at each turn on the twelve upper spirals; at the bottom of the tube the same reduction is carried out on the last eight spirals. The end of the ribbon is firmly tightened with an asbestos string; the two extremities are coated with a cement made of kaolin and sodium silicate (technical grade).

Trails of cement (width = 0.8 cm) are also drawn along three generatrices of the glass cylinder to maintain the ribbon which dilates when heated.

This assembly is dried for one day at room temperature and it is then introduced into another Pyrex tube (I.D. $\simeq$ 9 cm) of the same length. A small amount of cement is placed at each extremity. Then the external tube is coated with a sheet of alumina-silica felt WRP (Refractory Products Co., USP 3.092

247) impregnated with a suitable hardener. The thickness of the felt sheet is about four centimeters and it has been found best to place turns of thick paper between the glass tube and the insulator coating. This ensemble is slowly dried for one day at room temperature and one day in an oven at 100°C and is then progressively heated up to 500°C.

With a fine saw, two windows are cut in the alumina coating at the cell level to illuminate and reveal the electrodes. When the required materials are available, the construction of the furnace takes about three hours and, taking into account the drying periods, it is ready to operate after about four days.

Some improvements or modifications can be made for experimental purposes. Sometimes an additional heating source is introduced at the upper part of the tube to achieve a better homogeneity of temperature and to compensate for the heat loss through the electrode supports. This is carried out by six or seven spirals of heating ribbon wound on the external tube; the heating power depends on the body being heated and it is necessary to adjust the current with a power rheostat or, better still, with a variable transformer.

This type of furnace can be used at higher temperatures but the Pyrex tube has to be replaced. To retain the advantage of a transparent furnace, silica or Vycor tubing can be used. At a temperature above 700°C the Ni-Cr resistor gives a green deposit on the glass and it is better to use a platinum ribbon. The spirals are maintained by quartz pearls welded on the internal tube. As the platinum resistance increases several times with increasing temperature, a progressive increase of the voltage is required to reach a given temperature. An automatic adjustment of current can be used and a simple variable transformer has been suggested.[297] The slow-speed drive to the transformer is controlled by the heating current itself, but as the platinum may react with silica to form silicon-platinum alloy the maximum temperature should not exceed 850°C.

These transparent furnaces have a low thermal inertia. For example, with a silica cell containing about 200 or 300 g of fused electrolyte, a stable temperature at about 800°C can be obtained in less than one and a half hours, but this type of furnace is not recommended with an "on and off" regulating system which may give strong temperature oscillations.

When a large thermal inertia is required or when the working temperature is higher than 850°C, opaque refractory tubes are used; these furnaces are readily available commercially or they may be built in a few hours in the laboratory workshop. Spirals of wire resistor are placed round a supporting cylinder wound with thick paper; the winding is coated with a high alumina refractory cement and the cracks which often appear after drying at a low temperature ($\simeq$100°C) are filled with more cement. The cement piece is introduced into the furnace casing (iron tube) which is filled with an insulator material. Special care is taken with the electrical contacts between the resistor wire and the electrical leads and the resistance wire is doubly or triply twisted and fastened to the terminal block with screw contacts. Then a low heat is used to char the paper and sometimes a thin coat of cement is applied to cover the heating wire. The working temperature of such furnaces is limited by the maximum service temperature of the resistor.

Below 1000°C, the nickel–chrome (80%–20%) alloy is the best material; Kanthal* wire can be used up to 1200°C, but it becomes brittle when used and direct contact with SiO_2 must be avoided. For higher temperatures, we recommend platinum or platinum-rhodium alloy which can be used in an oxidizing atmosphere (air) with silica-free refractory material such as fused or recrystallized alumina. (Other resistance elements have been proposed such as silicon carbide resistors, but they are sensitive to humidity; long rods should be placed vertically to prevent deformation.) At low temperatures, the resistance of these elements decreases with temperature increase. As the elements show a gradual increase in resistance during use they should preferably be connected in parallel and since the resistance of each element is about 1 or 2 ohms a high current transformer is required. Other resistance heating elements, such as molybdenum or granular carbon particles, have been proposed to reach very high temperatures but they require inert or reducing atmospheres and are not much used in fused salt electrochemistry. A newer design, based on a precious metal film-coated tube is rapidly gaining popularity. Although expensive and generally only available from U.S. suppliers, the semi-transparent tube provides complete visibility of the experimental cell; heat uniformity is excellent and heat loss is minimal.

7.1.2. Insulation

The most common form of construction employs a refractory tube placed in a suitable casing of refractory insulation bricks or powder. Bricks can be easily shaped and they can be used to support resistors, by holes or slots cut in them. When powder is used a uniform small particle size will give the best insulating qualities; for porous bricks the smallest pore size gives the best insulation. The heat conductivity often shows a large temperature dependence and, in general, increases at high temperature.

The design of resistor-heated furnaces is mainly based on experience. However, when the thermal conductivity of the insulating material is known, it is possible to estimate the heat loss of the furnace. For a long tubular furnace with a concentric layer of insulation of effective conductivity k_m, the integration of the Fourrier law gives a heat loss q for a length l:

$$q = 2\pi k_m \frac{(T_1 - T_2)l}{\ln (d_2/d_1)}$$

where d_1 and d_2 represent the internal and external diameters of the cylindrical insulation, and T_1 and T_2 the temperature on these two surfaces; the surface temperature T_2 is quite hard to evaluate and depends on the surface emissivity

* Aktiebolaget Kanthal, Hallstahammar (Sweden): iron base alloy, 20–23 Cr, 4–5.5 Al, 0.6–1.0 Co, <0.1 C.

TABLE V. Properties of Insulating Materials

Material	Bulk density, g cm^{-3}	Max. service temperature, °C	Thermal conductivity, 10^{-4}J/cm S°C
Asbestos slab	0.132	600	8.62 at 200°C
Granular vermiculite	0.08–0.11	1100	10. 7 500°C
Vermiculite brick	0.48	1100	20.1 500°C
Diatomite powder	0.29	1200	8.62 1200°C
Insulating brick	0.79	1500	38. 1000°C
High alumina brick	2.72	1800	221. 500°C

and the surrounding environment. For a tubular furnace in still air the problem has been solved by Motzfeldt[291] who proposes a graphic approach to calculate the power input required by a resistance furnace in order to reach a given temperature.

The properties of different insulating materials are given in Table V.[298]

At the maximum temperature indicated the electrical resistance of these materials is in general good enough to avoid short-circuits between the heating elements. However, when a metallic cell is used, it happens that one electrode is in electrical contact with the external shield. In this case, the cell has to be carefully isolated from the furnace. This is very important when high impedance potentiostats are used; for example, in the cryolite experiment[217] at 1000°C we have observed that alumino-silicate bricks are not suitable to support the metallic cell; a coating of pure silico-alumina felt has to be employed to obtain a suitable response of the potentiostat.

ACKNOWLEDGEMENTS

Dr R. Narayan of the Department of Chemistry, Indian Institute of Technology, Madras, India, was the co-author of an unpublished article "Emf measurements in ionic melts at high temperatures", prepared some years ago and used as the basis of part of this chapter. We gladly acknowledge his contribution. Dr P. Dudley, Imperial College, assisted in coordination of this chapter.

References

1. P. Duby, *J Met.* **32,** 21 (1980).
2. M.J. Barbier, *Techniques de l'Ingénieur,* **J 3-690** (1966).
3. D. Inman and S.H. White, *J. Appl. Electrochem.* **8,** 375 (1978).
4.a M. Broc, G. Chauvin, and H. Coriou, in *Molten Salt Electrolysis in Metal Production,* Inter. Symposium at Grenoble, The Institution of Mining and Metallurgy, 44 Portland Place, London (1977), p. 69.
b V. Daňek and K. Matiašovský, *Surf. Technol.* **5,** 65 (1977).
5.a S.J. Yosim and K.M. Barclay, *Am. Chem. Soc., Div. Fuel Chem., Prepr.* **21,** 73 (1976).
b B. Ziólek and A. Bogacz, *Proceedings of the Third International Conference on Molten Salt Chemistry,* Wroclaw Karpacz, Poland, May-June (1979), p. 357.

6. "High Temperature Systems" in *Materials for Advanced Batteries,* D.W. Murphy, J. Broadhead, and B.C.H. Steele, Eds., Plenum Press, New York (1980), p. 315.
7. J.A. Plambeck, "Fused Salt Systems" in Vol. X, *Encylopedia of Electrochemistry of the Elements,* A.J. Bard, Ed., Marcel Dekker, New York (1976).
8. J.B. Headridge, *Electrochemical Techniques for Inorganic Chemists,* Academic Press, London (1969), p. 99.
9. J.W.E. Lewis and K. Singer, *J. Chem. Soc., Faraday Trans.* II **71**, 41 (1975).
10. L.V. Woodcock, *Advances in Molten Salt Chemistry,* Vol. 3, Plenum Press, New York (1975).
11. K. Kawamura and I. Okada, *At. Energy Rev.* **16,** 209 (1978).
12. F. Lantelme, P. Turq and P. Schofield, *J. Chem. Phys.* **71,** 2507 (1979).
13. D.D. MacDonald, *Transient Techniques in Electrochemistry,* Plenum Press, New York (1977).
14. M. Chemla and F. Lantelme, *Techniques de l'Ingénieur,* p. 2175 (1981).
15. F. Lantelme, J.P. Hanselin, and M. Chemla, *J. Electroanal. Chem.* **97,** 49 (1979).
16. R.S. Feigelson, *Adv. Chem. Ser.* **186,** 243 (1980).
17. F. Lantelme and J. Chevalet, *J. Electroanal. Chem.* **121,** 311 (1981).
18. F. Lantelme, D. Damianacos, and J. Chevalet, *Electrochem. Acta* **23,** 717 (1978).
19. A. Kisza and G. Zabinska, *Bull. Acad. Sci.*, Poland **25,** 473 (1977).
20. F. Lantelme, S. Belaidouni, *Electrochim. Acta* **26,** 1225 (1981).
21. K. Grjotheim, C. Krohn, M. Malinovský, K. Matiasŏvský, and J. Thonstad, *Aluminum Electrolysis,* Aluminum-Verlag GmbH, Dusseldorf, Germany (1977).
22. H. Ikeuchi and Y. Arita, Ext. Abstract, 30th Meeting of the I.S.E, Trondheim, Norway, August (1979), p. 28.
23. E.H. Amstein and W.D. Davis, p. 62 of Ref. 4a above.
24. Yu K. Delimarskii and B.F. Markov, *Electrochemistry of Fused Salts,* Sigma Press, Washington, D.C. (1961).
25. See, for example, N. Nachtrieb and M. Steinberg, *J. Am. Chem. Soc.* **72,** 3558 (1950); D. Inman, D.G. Lovering, and R. Narayan, *Trans. Faraday Soc.* **63,** 3017 (1967).
26. K.W. Fung and G. Mamantov, in *Advances in Molten Salt Chemistry,* Vol. 2, J. Braunstein, G. Mamantov, and G.P. Smith, Eds., Plenum Press, New York (1973).
27. *Metals Reference Book,* 5th Edition, J. Smithels, Ed., Butterworths, London (1976).
28. K. Kitazawa, S. Asakura, K. Fueki, and T. Mukaibo, *J. Electrochem Soc.* JPN, **37,** 45 (1969).
29. A.D. Graves, Thesis, London University (1967).
30. J.J. Egan and R. Heus, *J. Electrochem. Soc.* **107,** 824 (1960).
31. G.A. Wolstenholme, in *Molten Salt Technology,* D.G. Lovering, Ed., Plenum Press, New York (1982).
32. W.E. Cowley, in *Molten Salt Technology,* D.G. Lovering, Ed., Plenum Press, New York (1982).
33. W.W. Warren, Jr., in *Advances in Molten Salt Chemistry,* Vol 4, G. Mamantov and J. Braunstein, Eds., Plenum Press, New York (1981).
34. See, for example, S.K. Preto, Z. Tomczuk, S. von Winbush, and M.E. Roche, *J. Electrochem. Soc.* **130,** 264 (1983).
35. For a detailed discussion, see D.G. Lovering, Thesis, London University (1967).
36. J.J. Lingane, *Anal. Chim. Acta* **44,** 411 (1969).
37. D. Inman, D.G. Lovering, and R. Narayan, *Trans. Faraday Soc.* **63,** 3017 (1967).
38. R.M. Oblath, Thesis, CNNA (1978).
39. D. Inman, unpublished design.
40. H. Hoff, *Electrochim. Acta* **16,** 1059 (1971).
41. See, for example, J.E.B. Randles and W. White, *Z. Electrochem.* **59,** 666 (1955).
42. A.D. Graves and D. Inman, in *EMF Measurements in High Temperature Systems,* IMM Special Publications, C.B. Alcock, Ed. (1968).
43. For detailed discussion, see D.G. Lovering, Thesis, City University (1969).

44. G.J. Hills and P.D. Power, *J. Pol. Soc.* **13,** 71 (1967).
45. Yu. S. Lyalikov and V. Karmazin, *Zavod. Lab.* **14,** 144 (1948).
46. See, for example, G.J. Hills, J.E. Oxley, and D.W. Turner, *Silic. Ind.* **26,** 559 (1961); K.W. Fung and G. Mamantov, in *Comprehensive Analytical Chemistry,* C.L. Wilson and D.W. Wilson, Eds., Elsevier, New York (1975).
47. R. Narayan and D. Inman, *J. Pol. Soc.* **11,** 27 (1965).
48. K.E. Johnson, *J. Corr. Sci.* **6,** 87 (1967).
49. D.G. Lovering, Commercial report to NRDC/Rolls-Royce (1970).
50. P.G. Zambonin, F. Paniccia, and A. Bufo, *J. Phys. Chem.* **76,** 422 (1972).
51. J. Phillips, R.J. Gale, R.G. Wier, and R.A. Osteryoung, *Anal. Chem.* **48,** 1266 (1976).
52. See, for example, H.A. Laitinen, *Pure Appl. Chem.* **15,** 227 (1967); H.C. Gaur and H.J. Jindal, *J. Electroanal. Chem.* **16,** 437 (1967); *Indian J. Chem.* **5,** 624 (1967).
53. For an appraisal, see R.J. Gale, in *Molten Salt Technology,* D.G. Lovering, Ed., Plenum Press, New York (1982).
54. A.F. Alabyshev, M.F. Lantratov, and A.G. Morachevskii, *Reference Electrodes for Fused Salts,* Translated by A. Peiperl, Sigma Press, Washington, D.C. (1965); Yu. K. Delimarskii and B.F. Markov, *Electrochemistry of Fused Salts,* Translated by A. Peiperl, Sigma Press, Washington, D.C. (1961); R.W. Laity, in *Reference Electrodes: Theory and Practice,* D.J.G. Ives and G.J. Janz, Eds., Academic Press, New York (1961).
55. A.M. Shams El Din and A.A.A. El Hosary, *J. Electroanal. Chem.* **7,** 464 (1964); D. Inman and N.S. Wrench, *Br. Corros. J.* **1,** 246 (1966).
56. R. Littlewood, *Electrochim. Acta* **3,** 270 (1961).
57. See also Reference 276. J.P. Miller and R.P. Seward, *J. Phys. Chem.* **69,** 3156 (1965); A.J. Arvia, A.J. Calandra, and H.A. Videla, *Electrochim. Acta* **10,** 33 (1965). H.A. Videla and A.J. Arvia, *Electrochim. Acta* **10,** 21 (1965).
58. N.V. Bogoyavlenskaya, V.J. Chenenko, V.A. Babchenko, and E.J. Vydia, *Ukr. Khim. Zh.* **31,** 793 (1965).
59. H.A. Laitinen and J.A. Plambeck, *J. Am. Chem. Soc.* **87,** 1202 (1965). E.A. Ukshe and V.N. Devyatkin, *Electrokhimiya.* **1,** 627 (1965).
60. S. Flengas, *J. Chem. Soc.* 534 (1965). G.B. Shilina and Yu. K. Delimarskii, *Ukr. Khim. Zh.* **30,** 1045 (1964).
61. A.D. Graves and D. Inman, *Nature* **208,** 481 (1965).
62. E. Black and T.De. Vries, *Anal. Chem.* **27,** 906 (1955). Yu. K. Delimarskii and I.D. Panchenko, *Ukr. Khim. Zh.* **19,** 47 (1953); R. Narayan and D. Inman, *J. Pol. Soc.* **11,** 27 (1965).
63. G. Mamantov, in *Molten Salts: Characterization and Analysis,* G. Mamantov, Ed., Marcel Dekker, New York (1969), p. 537.
64. For example, M.F. Lantratov and A.F. Alabyshev, *Sh. Prikl. Khim.* (Moscow) **27,** 722 (1954); K.H. Stern, *J. Phys. Chem.* **62,** 385 (1958); S. Senderoff and G.M. Mellors, *J. Electrochem. Soc.* **105,** 224 (1958); R. Combes, J. Vedel, and B. Tremillon, *J. Electroanal. Chem.* **27,** 174 (1970); see also Ref. 54.
65. M.F. Lantratov and T.N. Shevlyakov, *Zh. Neorg. Khim.* **4,** 1153 (1959); **6,** 193 (1961); Yu. K. Delimarskii and A.A. Kolotii, *Ukr. Khim. Zh.* **14,** 124 (1948); H.A. Laitinen and J. Pankey, *J. Am. Chem. Soc.* **81,** 1053 (1959).
66. I.G. Murgulescu and L. Popescu, *Rev. Roum. Chim.* **17,** 1287 (1972).
67. G.J. Janz and F. Seagusa, *Electrochim. Acta* **7,** 293 (1962); Yu.K. Delimarskii and V.N. Andreeva, *Zh. Neorg. Khim.* **5,** 1123, 1800 and 2070 (1960); R. Sridhar and J.H.E. Jeffes, *Trans. I. Min. and Met.* **76,** C44 (1967).
68. N.S. Wrench, Ph.D. thesis, London (1967); W.E. Triaca and A.J. Arvia, *Electrochim. Acta* **9,** 1055 (1964).
69. See D. Inman and N.S. Wrench, in Ref. 55.
70. J.R. Selman and H.C. Maru, in *Advances in Molten Salt Chemistry,* Vol. 4, G. Mamantov and J. Braunstein, Eds., Plenum Press, New York (1981).

71. H.A. Laitinen and B.B. Bhatia, *J. Electrochem. Soc.* **107,** 705 (1960); S. Karpachev, S.I. Rempel, and E. Iordan, *Zh. Fiz. Khim.* **23,** 422 (1949).
72. M. Takahashi, *Denki Kagaku* **25,** 481 (1957); L.E. Ivanoskii and M.T. Krasil'nikov, *Trudy in-ta Electrokhimii UFAN USSR,* Issue 2, 79 (1961); B.P. Artamonov, *Trudy G I P Khim.,* Issue 33, 39 (1940); and see Alabyshev *et al.,* Ref. 54.
73. J. Corbett and S.v. Winbush, *J. Am. Chem. Soc.* **77,** 3964 (1955). S. Karpachev, A. Stromberg, and E. Iordan, *Zh. Fiz. Khim.* **18,** 43 (1944); but see J.W. Johnson and M.A. Bredig, *J. Phys. Chem.* **62,** 604 (1958).
74. Yu.K. Delimarskii, *Zh. Fiz. Khim.* **24,** 875 (1950); **29,** 28 (1955); and see also Refs. 54 and 78.
75. V.A. Pleskov, *Usp. Khim.* **16,** 254 (1947); Idem, *J. Phys. Chem.* (USSR) **22,** 351 (1948), but see H. Schnreider and H. Strehlow, *J. Electroanal. Chem.* **12,** 530 (1966).
76. M.A. Bredig and H.R. Bronstein, *J. Phys. Chem.* **64,** 46 (1960); M.A. Bredig, J.W. Johnson, and W.T. Smith, *J. Am. Chem. Soc.* **77,** 307 (1955); M.A. Bredig, in *Molten Salt Chemistry,* M. Blander, Ed., Interscience, New York, (1964).
77. K. Hauffe, *Z. Electrochem.* **46,** 348 (1940).
78. Yu.K. Delimarskii *et al., Ukr. Khim. Zh.* **15,** 77 (1949); **16,** 1 (1950); **20,** 502 (1954). Idem, *Zh. Fiz. Khim.* **28,** 1169 (1954).
79. Yu.K. Delimarskii and A.A. Kolotii, *Ukr. Khim. Zh.* **24,** 146 (1958).
80. K.H. Stern, *Chem. Rev.* **66,** 355 (1966).
81. H.A. Laitinen and C.H. Liu, *J. Am. Chem. Soc.* **80,** 1015 (1958).
82. For example, J. Braunstein and M. Blander, *J. Phys. Chem.* **64,** 10 (1960); R.J. Labrie and V.A. Lamb, *J. Electrochem. Soc.* **106,** 875 (1959); and see references listed under 90.
83. E.L. Colichman, *Anal. Chem.* **27,** 1559 (1955).
84. M. Steinberg and N.H. Nachtriev, *J. Am. Chem. Soc.* **72,** 3558 (1950).
85. F.R. Duke and H.M. Garfinkel, *J. Phys. Chem.* **65,** 1627 (1961).
86. M. Francini and S. Martini, *EURATOM 1908.e* (1964).
87. J. Braunstein and A.S. Minano, *Inorg. Chem.* **5,** 942 (1966). M. Francini and S. Martini, in *Polography 1964,* G.J. Hills, Ed., Macmillan, London (1966).
88. See H.A. Laitinen and B.B. Bhatia in Ref. 71.
89. D. Inman, *Electrochim. Acta* **10,** 11 (1965).
90. J.O'M. Bockris, G.J. Hills, D. Inman, and L. Young, *J. Sci. Instrum.* **33,** 438 (1956); D. Inman, *J. Sci. Instrum.* **39,** 391 (1962); K.H. Stern and S.E. Meader, *J. Res. Nat. Bur. Stand.* **69A,** 553 (1965), and previous work cited therein.
91. See R.J. Labrie and V.A. Lamb, in Ref. 82; D. Inman, in Ref. 90.
92. For example, G. Grube and E.A. Rau, *Z. Electrochem.* **40,** 352 (1934); Yu.K. Delimarskii and A.A. Kolotii, *Zavod. Lab.* **22,** 25 (1956); A.A. Kolotii, *Ukr. Khim. Zh.* **24,** 572 (1958); W.E. Reid, *J. Electrochem. Soc.* **108,** 393 (1961); R.J. Labrie and V.A. Lamb, in Ref. 82.
93. H. Coriou, J. Dirian, and J. Hure, *J. Chim. Phys.* **52,** 479 (1955); V.J. Minenko, N.S. Ivanova, and J.K. Falko, *Ukr. Khim. Zh.* **31,** 804 (1965); G. Mamantov, D.L. Manning, and J.Dale, *J. Electroanal. Chem.* **9,** 253 (1965).
94. J.P. Tverdovskii and V.S. Molchanov, *J. Appl. Chem. (USSR)* **10,** 1011 (1937); J. Besson *et al. Compt. rend.* **260,** 2497 (1965); and see Ref. 106.
95. See H. Coriou *et al.,* Ref. 93.
96. H.J. Hein, *Chem. Ing. Tech.* **32,** 695 (1960); see also Refs. 85, 86.
97. S. Senderoff and A. Brenner, *J. Electrochem. Soc.* **101,** 16, 28, 31 (1954); **102,** 77 (1955).
98. For example, J. Braunstein and A.S. Minano, *Inorg. Chem.* **3,** 218 (1964).
99. M. Bonnemay and R. Pineaux, *Compt. rend.* **240,** 1774 (1955).
100. D.G. Hill and M. Blander, *J. Phys. Chem.* **65,** 1866 (1961).
101. R. Littlewood, *Electrochim. Acta* **3,** 270 (1961); E.A. Ukshe, N.G. Bukun, D.I. Leikis, and A.N. Frumkin, *Electrochim. Acta* **9,** 431 (1964); B.W. Burrows, Ph.D Thesis, Southampton University (1965).

102. M. Petit, *Compt. rend.* **258,** 6143 (1964); **261,** 1301 (1965).
103. M. Rey, G. Danner, and M. Abraham, *Compt. rend.* **248,** 2868 (1959); and see Ref. 102.
104. R.J. Labrie and V.A. Lamb, *J. Electrochem. Soc.* **106,** 875 (1959).
105. Unpublished work in D. Inman's laboratory by E. Franks and D.M. Wrench.
106. J. Besson, C. Deportes, and M. Darcy, *Compt. rend.* **251,** 1630 (1960); M.V. Perfil'ev, S.F. Pal'guev, and S.V.K. Karpachev, *Sov. Electrochem.* **1,** 74 (1965); but see M.F. Lantratov and A.G. Morachevskii, *Zh. Prikl. Khim.* **32,** 1397 (1959); A. Mitchell, *Trans. Faraday Soc.* **62,** 3470 (1966).
107. S. Senderoff, G.W. Mellors, and W.J. Reinhart, *J. Electrochem. Soc.* **112,** 840 (1965).
108. R. Winand, *Electrochim. Acta* **17,** 251 (1972).
109. J. Thonstad, *J. Electrochem. Soc.* **10,** 346 (1968).
110. S. Vire, Ph.D Thesis, London (1981).
111. H.R. Bronstein and D.L. Manning, *J. Electrochem. Soc.* **19,** 125 (1972).
112. F.R. Clayton, G. Mamantov, and D.L. Manning, *J. Electrochem. Soc.* **120,** 1153 (1973).
113. F. Lantelme, S. Belaidouni, and M. Chemla, *J. Chim. Phys.* **76,** 423 (1979).
114. R. Knödler, G. Bohme, and W. Borger, *J. Electrochem. Soc.* **127,** 1311 (1980).
115. M.D. Ingram, G.G. Lewis, and J.A. Duffy, *J. Phys. Chem.* **76,** 1035 (1972).
116. G.W. Mellors and S. Senderoff, *J. Electrochem. Soc.* **112,** 266 (1965).
117. D. Damianacos, F. Lantelme, and M. Chemla, *Electrochim. Acta* **26,** 917 (1981).
118. S.S. Barton, G.L. Bouton, and B.H. Harrison, *Carbon* **10,** 395 (1972).
119. C.E. Bamberger, in *Advances in Molten Salt Chemistry,* Vol. 3, J. Braunstein, G. Mamantov, and G.P. Smith, Eds., Plenum Press, New York (1975).
120. J. Goret, *Bull. Soc. Chim. Fr.* 1074 (1964).
121. D.R. Flinn and K.H. Stern, *J. Electroanal. Chem.* **63,** 39 (1975).
122. E.R. Buckle and D.E. Dendrinos, *Proceedings of the International Symposium on Molten Salts,* J.P. Pemsler *et al.*, Ed., The Electrochemical Society, Pennington, New Jersey (1976), p. 494.
123. F. Lantelme and M. Chemla, *Bull. Soc. Chim. Fr.* 762 (1974).
124. R. Szarc and R.D. Walton, Ext. Abstracts of the 30th Meeting of I.S.E., Trondheim, Norway (1979), p. 60.
125. G.O. Nelson, *Controlled Test Atmospheres., Principles & Techniques,* Science, Ann Arbor, Michigan (1971).
126. Catalogue Gaz Material, p. 223 (1980), L'Air Liquide Division Scientifique Alphagaz, 75 quai d'Orsay 75231, PARIS CEDEX 07.
127. O. Grubner, P. Jirú, and M. Rálek, "Molekularsiebe," in *Physikalische-Chemische Trennung Methoden,* (Bd 12), VEB Deutscher Verlag der Wissenschaften, Berlin (1968).
128. P. Dunsheath, B.P 479, 914 14-2 (1938).
129. A.S. Newton, U.S.P 2, 487, 360, 8-11 (1949).
130. A.D. Kelmers and M.R. Bennett, p. 406 of Ref. 122 (1976).
131. M.W. Mallett, *J. Ind. Eng. Chem.* **42,** 2095 (1950).
132. F. Maak and C.M. Sellars, *J. Sci. Instrum.* **42,** 900 (1965).
133. Matheson, *Cie, Industrial Research and Development* **23,** 24 (1981).
134. G. Boisde, C. Chauvin, and H. Coriou, *Electrochim. Acta* **11,** 375 (1966).
135. R.J. Gale and R.A. Osteryoung, *J. Electrochem. Soc.* **121,** 993 (1974).
136. M. Souders, Canadian Patent 437 940 12.11 (1946); see also Ref. 148.
137. D.L. Maricle and D.N. Hume, *J. Electrochem. Soc.* **107,** 354 (1966).
138. H.A. Laitinen and J.W. Pankey, *J. Am. Chem. Soc.* **81,** 1053 (1959).
139. W.G. Burns and F.S. Dainton, *Trans. Faraday Soc.* **48,** 39 (1952).
140. O.P. Mehta, F. Lantelme, and M. Chemla, *Rev. Int. Hautes Temper. et Refract.* **4,** 21 (1967).
141. M. Chemla, French Patent No. 1 216 418, 25 April (1960).
142. C.F. Baes, Jr. in *Symposium on Reprocessing of Nuclear Fuels, Nuclear Metallurgy,* Vol. 15, P. Chiotti, Ed., CONF-690801 (1969), p. 615.
143. M. Caron, P. Coste, C. Coquet, and M. Rey, French Patent No. 70 08897 (1970).

144. M.E. Runner, G. Baldog, and M. Kilpatrick, *J. Am. Chem. Soc.* **78,** 5183 (1956).
145. A.T. Kuhn, *Industrial Electrochemical Processes,* Elsevier, London (1971), p. 7.
146. D. Devilliers, F. Lantelme, and M. Chemla, *J. Chim. Phys.* **76,** 428 (1979).
147. S. Aoyama and E. Kanda, *J. Chem. Soc. Jpn* **58,** 706 (1937).
148. H.A. Laitinen and J.A. Plambeck, *J. Amer. Chem. Soc.* **87,** 1202 (1965).
149. H. Melville and B.G. Gowenlock, *Experimental Methods in Gas Reactions,* MacMillan, New York (1964).
150. H.E. Bartlett and K.E. Johnson, *J. Electrochem. Soc.* **114,** 457 (1967).
151. J. Dubois, *Ann. Chim. (Paris)* **10,** 145 (1965).
152. H.A. Liebhafsky and E.J. Cairns, *Fuel Cells and Fuel Batteries,* J. Wiley and Sons, New York (1968).
153. A. Borucka, *Electrochim. Acta* **13,** 295 (1968).
154. R.H. Arendt and M.J. Curran, *J. Electrochem. Soc.* **127,** 1663 (1980); see also Ref. 259.
155. L.G. Boxall and K.E. Johnson, *J. Electrochem. Soc.* **118,** 885 (1971).
156. P.G. McCormick and H.S. Swofford, Jr., *Anal. Chem.* **41,** 146 (1969).
157. F. Lantelme, Thesis, University of Paris (1965); C.E.A. Report R. 2806 (1965).
158. Y. Kanzaki and M. Takahashi, *J. Electroanal. Chem.* **58,** 339 (1975).
159. J. Besson, C. Deportes, and P. Donneaud, *Comptes Rendus Acad. Sci. Paris* **260,** 2497 (1965).
160. Y.I. Krotov and Ya.E. Vil'Nyanskii, *Zh. Prikl. Khim.* **38,** 1206 (1965).
161. M. Pourbaix, Ext. Abstract, 30th Meeting of I.S.E at Trondheim, Norway, id. Ref. 22, p. 275 (1979).
162. Yu.K. Delimarskii and N.K. Tumanova, *Ukr. Khim. Zh.* **34,** 988 (1968).
163. A. Lunden, Thesis, Gothenburg, Sweden (1956).
164. I. Okada, K. Okuyama, T. Miyamoto, I. Tomita, and N. Saito, *J. Inorg. Nucl. Chem.* **35,** 2957 (1973).
165. R.W. Graham, *Secondary Batteries, Recent Advances,* Noyes Data Corporation, Park Ridge, New Jersey (1978).
166. S.N. Flengas and A. Block-Bolten, *Advances in Molten Salt Chemistry,* Vol. 2 (1973), p. 27.
167. D.G. Lovering and R.M. Oblath, *J. Electrochem. Soc.* **127,** 1997 (1980).
168. K.H. Stern, Rm. Panayappan, and D.R. Flinn, *J. Electrochem. Soc.* **124,** 641 (1977).
169. R.E. Gibson and L.H. Adams, *J. Am. Chem. Soc.* **55,** 2679 (1933).
170. T.B. Tripp and J. Braunstein, *J. Phys. Chem.* **73,** 1984 (1969).
171. A. Fontana and R. Winand, *Electrochim. Acta* **16,** 257 (1971).
172. D. Damianacos, F. Lantelme, and M. Chemla, *Electrochim. Acta* **28,** 217 (1983); see also Ref. 18.
173. A. Borucka and C.M. Sugiyama, *Electrochim. Acta* **13,** 1887 (1968).
174. A. Bonomi, M. Hadate, and C. Gentaz, *J. Electrochem. Soc.* **124,** 982 (1977).
175. G.H. Miller, *Technique of Organic Chemistry,* Vol. III, Weissberger, Ed., Interscience Publishers, New York (1957), Part II.
176. B. Cleaver, *J. Chem. Soc., Faraday Trans. I* **75,** 2735 (1979).
177. K. Tödheide, id. Ref. 122, p. 20 (1976).
178. S.D. Snyder, C.J. Craven, and D.B. Trauger, *Rev. Sci. Instrum.* **21,** 887 (1950).
179. A.C. Burton and J.T. Nichol, *Rev. Sci. Instrum.* **21,** 485 (1950).
180. W.S. Peterson, *Ind. Eng., Chem. Anal. Ed.* **17,** 54 (1945).
181. L. Domange and J. Neudorfer, *Bull. Soc. Chim. Fr.* 728 (1949).
182. R.M. Edge, *J. Sci. Instrum. (J. Phys. E.)* **1** (2), 35 (1968).
183. P. Bradshaw, *J. Sci. Instrum. (J. Phys. E.)* **1** (2) 504 (1968).
184. J.A. Vepřek, *J. Sci. Instrum.* **40,** 66 (1963).
185. W. Snelleman, *J. Sci. Instrum.* **43,** 832 (1966).
186. G.G. Oberfell and R.P. Mase, *Ind. Eng. Chem.* **11,** 294 (1919).
187. H. Wosthoff Ohg., *Digamix Gas Mixing Pumps,* D. 4630 BOCHUM 1, Production Programm 511e (1980).
188. J.A.A. Ketelaar and A. Dammers de Klerk, *Rec. Trav. Chim. Pays-Bas* **83,** 322 (1964).

189. P.G. Zambonin, *J. Electroanal. Chem.* **33,** 243 (1971).
190. B.E. Saltzman, *Anal. Chem.* **33,** 1100 (1961).
191. M. Maman, G. Valdener, and R. Massot, in *Précis de Spectrométrie de Masse Analytique,* Presse Universitaire de Grenoble, France (1975).
192. J.C. Touray and F. Lantelme, *C.R. Acad. Sci. Paris* **262,** 1 (1966); see also Ref. 246.
193. A. Cornu, and R. Massot, *Compilation of Mass Spectral Data,* Heyden, London (1975).
194. L. Spzepesy, *Gas Chromatography,* Engl. Trans., Iliffe Books Ltd., London (1970).
195. M. Lucas, and G. Frejaville, in *Précis de Spectrométrie de Masse Analytique,* A. Cornu, Ed., Presse Universitaire de Grenoble, France (1975).
196. G.E. Baiulescu and V.A. Ilie, *Stationary Phase in Gas Chromatography,* Pergamon Press, Oxford (1975).
197. N.L. Alpert, W.E. Keiser, and H.A. Szymanski, *Theory and Practice of Infrared Spectroscopy,* Plenum Press, New York (1970).
198. G. Herzberg, *Molecular Spectra and Molecular Structure,* D. Van Nostrand Company, New York (1960).
199. N. Watanabe and T. Asave, Denki Kagaku **39,** 123 (1971).
200. M. Bergeret, *Energ. Nucl. (Paris),* Mars-Avril, 93 (1965).
201. G. Jessop, *J. Sci. Instrum.* **43,** 777 (1966).
202. G. Bianchi, G. Faita, and T. Mussini, *J. Sci. Instrum.* **42,** 693 (1965).
203. W.R.S. Garton, M.S.W. Webb, and P.C. Wildy, *J. Sci. Instrum.* **34,** 496 (1957).
204. R.D. Goodwin, *Anal. Chem.* **25,** 263 (1953).
205. J.W. Ross, J.H. Riseman, and J.A. Kruger, *Pure. Appl. Chem.* **36,** 473 (1973).
206. A. Dolegeal, D. Devilliers, G. Villard, and M. Chemla, *Analusis* **10,** 377 (1982).
207. A. Hrabeczy-Pall, K. Toth, E. Pungor, and F. Vallo, *Anal. Chim. Acta* **77,** 278 (1975).
208. B.W. Burrows and G.J. Hills, *Electrochim. Acta* **15,** 445 (1970).
209. O.P. Mehta, F. Lantelme, and M. Chemla, *Electrochim. Acta* **14,** 505 (1969).
210. G.J. Janz, *Experimental Techniques in Molten Salts Handbook,* Academic Press, New York (1967).
211. F.G. Bodewig and J.A. Plambeck, *J. Electrochem. Soc.* **117,** 904 (1970).
212. O.A. Esin, B.M. Lepinskikh, and V.I. Musikhin, *Isvest. Akad. Nauk. S.S.S.R., Otdel. Tekh. Nauk.,* (12) 120 (1954).
213. A. Bonomi, G. Gentaz, M. Parodi, and J.L. Hermin, in *Molten Salt Electrolysis in Metal Production.* International Symposium at Grenoble, The Institution of Mining and Metallurgy, 44 Portland Place, London, W1N 4BR (1977), p. 14.
214. D. Damianacos, Thesis, Université P et M. Curie, Paris (1980); see also Ref. 117.
215. R.J. Ring and D. Royston, Aust. Atom. Energ. Com., Report AAEC/E-281 (1973).
216. N. Nardin, E. Chassaing, and G. Lorthioir, id. Ref. 213, p. 38 (1977).
217. F. Lantelme, D. Damianacos, and M. Chemla, *J. Electrochem. Soc.* **127,** 498 (1980).
218. D. Inman, G.J. Hills, L. Young, and J. O'M. Bockris, *Ann. N.Y. Acad. Sci.* **79,** 803 (1960).
219. M. Chemla and V. Grinevitch, *Bull. Soc. Chim. Fr.* 853 (1973).
220. B. Mazza, P. Pedeferri, and A. Tognoni, *Chim. e. Ind. (Milano)* **53,** 123 (1971).
221. W.E. Haupin and W.C. McGrew, in *Light Metals 1973,* 103rd AIME Meeting, Dallas, Texas (1973), p. 37.
222. L.M. Toth, J.P. Young, and G.P. Smith, *Anal. Chem.* **41,** 683 (1969).
223. B.W. Hatt, and D.H. Kerridge, *Chem. Br.* **15,** 78 (1979).
224. C.F. Baes, *J. Nucl. Mater.* **51,** 149 (1974).
225. D.O. Raleigh, J.T. White, and C.A. Ogden, *J. Electrochem. Soc.* **126,** 1087 (1979).
226. M. Roche, *Rev. Phys. Appl.* **15,** 895 (1980).
227. C.A. Melendres, J.P. Ackerman, and R.K. Steunenberg, *Proceedings of the International Symposium of Molten Salts,* J.P. Pemsler *et al.*, Ed., The Electrochemical Society, Inc., Pennington, New Jersey (1976), p. 575.

228. H. Imoto and N. Watanabe, *Bull. Chem. Soc. Jpn.* **49** (1976); H. Imoto, K. Ueno and N. Watanabe, id. **51,** 2822 (1978).
229. C.E. Bamberger, *Advances in Molten Salts Chemistry,* Vol. 3 (1975), p. 177.
230. A.T. Kuhn, *Encyclopedia of Electrochemistry of the Elements,* Vol. IV, Marcel Dekker, New York (1975).
231. J.R. Keiser, D.L. Manning, and R.E. Clausing, id. Ref. 227 (1976), p. 315.
232. C.E. Bamberger and C.F. Baes, Jr., *J. Am. Ceram. Soc.* **55,** 564 (1972)
233. H.N. Jenkins, G. Mamantov, and D.L. Manning, *J. Electroanal. Chem.* **19,** 385 (1968).
234. D. Damianacos, F. Lantelme, and M. Chemla, *J. Chim. Phys.* **76,** 391 (1979).
235. J. Thonstad, *Electrochim. Acta* **14,** 127 (1969).
236. K. Grjotheim, M. Malinovský, and K. Matiašovský, *J. Met.* **2,** 28 (1969).
237. J. Thonstad, *Electrochim. Acta* **6,** 346 (1968).
238. K. Grjotheim, M. Malinovský, M. Matiašovský, A. Silný, and J. Thonstad, *Can. Met. Quart.* **11,** 295 (1972).
239. P. Homsi, Thesis, Université Pierre et Marie Curie (1975).
240. H.A. Laitinen and J.A. Plambeck, *J. Am. Chem. Soc.* **87,** 1202 (1965).
241. J.A. Plambeck, "Fused Salt Systems," in *Encyclopedia of Electrochemistry of the Elements,* Vol. X, Marcel Dekker, New York (1976), p. 12.
242. M. Bruneaux, G. Darmois, and S. Ziolkiewicz, *C.R. Acad. Sci. Paris,* **C254,** 2967 (1962).
243. See "notice on vitreous carbon," 'Société Le Carbone Lorraine', 41 rue Jean-Jaurès, 92231 Gennevilliers, France.
244. J.H. Kennedy and F. Adamo, *J. Electrochem. Soc.* **119,** 1517 (1972).
245. D.R. Morris, C.A. Aksaranan, B.S. Waldron and S.H. White, *J. Electrochem. Soc.* **120,** 570 (1973); see also Ref. 174.
246. D. Devilliers, M. Vogler, F. Lantelme, and M. Chemla, *Anal. Chim. Acta* **153,** 69 (1983).
247. H.V. Venkatasetty and D.J. Saathoff, id. Ref. 227, p. 329 (1976).
248. M.M. Wong and F.P. Haver, id. Ref. 213, p. 21 (1977).
249. F. Lantelme and M. Chemla, *Electrochim. Acta* **10,** 663 (1965).
250. P.G. Zambonin and J. Jordan, *J. Am. Chem. Soc.* **91,** 2225 (1969)
251. D.R. Flinn and K.H. Stern, *J. Electroanal. Chem.* **63,** 39 (1975).
252. M. Fredericks, R.B. Temple, and G.W. Thickett, *J. Electroanal. Chem* **38,** App. 5 (1972).
253. K.E. Johnson, P.S. Zacharias, and J. Matthews, id. Ref. 227, p. 603 (1976).
254. E. Desimoni, F. Paniccia, L. Sabbatini, and P.G. Zambonin, id. Ref. 227, p. 584 (1976).
255. D.L. Douglas in *Fuel Cells,* Vol. 1, G.J. Young, Ed., Reinhold, New York (1960), p. 129.
256. A.J. Appleby and S. Nicholson, *J. Electroanal. Chem.* **53,** 105 (1974).
257. V.V. Kuzmovich and S.S. Fedorchenko, *Ukr. Khim. Zh.* **32,** 846 (1966).
258. L.F. Grantham and P.B. Ferry, id. Ref. 227, p. 270 (1976).
259. R.H. Arendt, K.W. Browall, and N.S. Choudhury, Ext. Abstracts of 30th Meeting of I.S.E., Trondheim, Norway (1979), p. 83.
260. K.E. Johnson and H.A. Laitinen, *J. Electrochem. Soc.* **110,** 314 (1963).
261. J. Goret, Thesis, Paris University (1966); see also Ref. 120.
262. J. Goret and B. Tremillon, *Bull. Soc. Chim. Fr.,* 67 (1966).
263. J.N. Agar and F.P. Bowden, *Proc. R. Soc. of London, Ser. A* **169,** 206 (1938).
264. K.H. Stern and J.K. Carlton, *J. Phys. Chem.* **38,** 965 (1954).
265. J. Goret and B. Tremillon, *Electrochim. Acta* **12,** 1065 (1967).
266. R. Lenke, W. Uebelhack, and A. Klemm, *Z. Naturforsch.* **28a,** 881 (1973).
267. D. Adams and G. Hills, in *Ionic Liquids,* D. Inman and D.G. Lovering, Eds., Plenum Press, New York (1981), p. 27; see also Ref. 10.
268. F. Lantelme, and P. Turq, *J. Chem. Phys.* **77,** 3177 (1982).
269. The text of this scale is published in Compt. Rend. 13th Conf. gén. Poids et Mesures, Annexe 2 (1967). The English text is published in *Metrologia.* **5,** 35 (1969).

270. J. A. Hall, *J. Sci. Instrum.* **44,** 257 (1967).
271. R.E. Bentley and T.P. Jones, *High Temp.-High Pressures* **12,** 33 (1980).
272. E.D. Zysk, in *Temperature, Its Measurement and Control in Science and Industry,* Vol. 3, C.M. Herzfeld, Ed., Reinhold, London (1962), Part 2, p. 135.
273. P. Mergault and C. Brunet-Lefort, *C.R. Acad. Sci. Paris.* **276 C,** 29 (1973).
274. C. Guilpin and J. Garbarz-Oliver, *Thermochim. Acta* **13,** 467 (1975).
275. T. Shikama, T. Tanabe, M. Fujitsuka, H. Yoshida, and R. Watanabe, *Tetsu To Hagane* **66,** 418 (1980).
276. *Temperature Measurement Handbook,* Omega Engineering, Inc., Box. 4047, Stamford, Connecticut 06907, USA (1983).
277. P. Malmberg, *Rev. Sci. Instrum.* **27,** 136 (1956).
278. H.C. Wolfe, in *Temperature, Its Measurement and Control in Science and Industry,* Reinhold, New York (1955), p. 288.
279. K. Bethe, in *Practical Measurement for Improving Efficiency,* Acta Imeko 1976, Vol. II., p. 499, Akademia Kiadó, Budapest, Hungary (1977).
280. *Temperature Handbook,* Nanmac Corporation, 9-11 Mayhew Street, Framingham Ctr., Mass. 01701, USA (1981).
281. J.A. Hall and V.M. Leaver, *J. Sci. Instrum.* **38,** 178 (1961).
282. E. Iwanciw, A. Warczok, J. Wojewoda, and W. Mroz, *Rudy. Met. Niezelaz.* **22,** 482 (1977).
283. R. Boen, Thesis, Institut National Polytechnique de Grenoble, Saint-Martin d'Hères, France (1979).
284. R. Winand, A. Fontana, L. Segers, P. Hammaert, and J. Lacave, id. Ref. 213, p. 42 (1977).
285. D.A. Davies, *J. Sci. Instrum.* **37,** 15 (1960).
286. H.J. Scheel and C.H. West, *J. Phys., E* **6,** 1178 (1973).
287. Application notes, VMOS Power Fet's, Siliconix Inc., Melville, L.I., NY 11746, USA (1979).
288. T.J. Maloney, *Industrial Solid-State Electronics, Devices and Systems,* Prentice-Hall, New Jersey (1979).
289. J.L. Cribb, *J. Sci. Instrum.* **39,** 128 (1962).
290. M. Cordier, *Transfert de chaleur par conductibilité,* INSTN, Centre de Documentation Universitaire, Paris (1961).
291. K. Motzfeldt, in *Physicochemical Measurements at High Temperatures,* J.O'M. Bockris, J.L. White and J.D. Mackenzie, Eds., Butterworths, London (1959).
292. R.W. Laity, in *Reference Electrodes,* D.J.G. Ives and G.J. Janz, Eds., Academic Press, New York (1961), p. 524.
293. R. Foulatier, M. Guerin, C. Vovelle, and H. Mellottée, *J. Phys. E.* **4,** 291 (1971).
294. M.B. Reynolds, *Rev. Sci. Instrum.* **25,** 838 (1954).
295. R. Severns, *Electronic Design.* **3,** 85 (1980).
296. A. de Guibert and V. Plichon, *J. Electroanal. Chem.* **90,** 399 (1978).
297. W.W. Marshall, *J. Sci. Instrum.* **42,** 663 (1965).
298. P.L. Start and M.W. Thring, *J. Sci. Instrum.* **37,** 17 (1960).

6

Automated Admittance Spectroscopy of the Semiconductor/Molten Salt Electrolyte Interface

Krishnan Rajeshwar

1. Introduction

The efficacy of semiconductor/electrolyte junctions and, in particular, semiconductor/molten salt interfaces for photovoltaic and photoelectrochemical applications is critically dependent on how well processes at the interface (e.g., charge transfer, carrier recombination, specific ion adsorption) can be controlled. Use of the term "interface" herein is made with the implicit recognition that the boundary between the two phases is not *sharp* but several Å thick. Development of measurement techniques and physical models for characterizing these interfacial processes, therefore, is of particular value in designing and optimizing an efficient solar conversion system. Equivalent circuit representations are useful in delineating each individual electrophysical (e.g., charge accumulation) or electrochemical (e.g., faradaic charge transfer) process in the system and their interrelationships in terms of an overall electrical network. Identification of individual circuit elements comprising the overall network is readily accomplished by impedance or admittance measurements. Such measurements have been traditionally employed in kinetics studies of electrochemical charge transfer across the electrode/electrolyte interface (*cf.* Refs. 1 and 2). Less attention has been devoted to electrophysical aspects of the interface. In this chapter, models will be developed whereby individual circuit elements comprising a semiconductor/electrolyte junction may be resolved by automated admittance measurements. The treatment is specific to the molten salt electrolyte case, although extrapolation to aqueous electrolyte situations and the general case of electrode/electrolyte interfaces is fairly straightforward.

Krishnan Rajeshwar • Department of Chemistry, University of Texas at Arlington, Arlington, Texas 76019, U.S.A.

2. Measurement Techniques

The electrical impedance or admittance of a given system is most conveniently described in terms of the generalized Ohm's law (Equation 1) provided that its response to any arbitrary electrical perturbation is linear.[3]

$$V(t) = Z(p) \cdot I(t) \tag{1}$$

$$\text{or } I(t) = Y(p) \cdot E(t) \tag{2}$$

where $V(t)$ and $I(t)$ are the voltage and current, respectively, both of which are functions of time, t. The terms $Z(p)$ and $Y(p)$ are the impedance and admittance functions, and the operator p is defined by Equation 3.

$$p = \mathrm{d}/\mathrm{d}t, \frac{1}{p} = \int \mathrm{d}t \tag{3}$$

The functions $Z(p)$ and $Y(p)$ operate on the independent variables $I(t)$ and $V(t)$, respectively, to yield the appropriate dependent variables. For example, the voltage drops across a single resistor and capacitor in response to an applied current perturbation are given by Equations 4 and 5, respectively:

$$V(t) = RI(t), Z(p) = R, Y(p) = \frac{1}{R} \tag{4}$$

$$V(t) = \frac{1}{C} \int I\mathrm{d}t, Z(p) = \frac{1}{pC}, Y(p) = pC \tag{5}$$

For the similar case of a voltage perturbation, the operator $Y(p)$ is chosen so that the response $I(t)$ is equal to the operation $Y(p) \cdot V(t)$. Analogous concepts apply to more complex circuits, such as the ones treated in this chapter, provided that the individual elements in the circuit are linear.

Consider again the application of a sine wave voltage of the form $V(t) = V_0 \sin \omega t$ where ω is the angular frequency ($2\pi f$) and V_0 is the peak amplitude of the signal, to a simple resistor. The current through the resistor will be a sine wave of the form, $I(t) = I_0 \sin \omega t$. In the case of a capacitance element, however, it is obvious that the potential across the capacitor cannot change instantaneously because an infinite current would be required. The resultant phase lag between the applied voltage and the resulting current is described in terms of a phase angle, θ, such that $I(t) = I_0 (\sin \omega t + \theta)$. In a pure capacitor, this angle will be 90°. Thus, ac circuits comprise both resistive as well as reactive components for which the impedance or admittance functions are conveniently described by complex notations (Equations 6 and 7 respectively):

$$Z(p) = R + jX \tag{6}$$

$$Y(p) = G + jB \tag{7}$$

In these equations, $j = \sqrt{-1}$, X is called the capacitive reactance, G is the conductance $\left(\frac{1}{R}\right)$, and B is termed the capacitive susceptance.

The electrical impedance (or admittance) of a solid/liquid interface has been measured by a variety of techniques.[3] These techniques may be broadly divided into those based on measurements either in the time domain or in the frequency domain. Both types of measurements are entirely complementary and contain identical information because of the interrelationship of time and frequency. Frequency domain measurements, however, have an advantage in that processes which exhibit a wide range of relaxation times (typical of semiconductor/solution interfaces) are accessible in a single experiment. Only frequency domain measurement techniques are treated in this chapter—the reader being referred to References 3 and 4 for a detailed discussion of time domain methods.

Frequency domain measurements may be further subdivided into ac bridge methods or null techniques.[5] In the latter, either phase-sensitive detection or direct determination of the in-phase and quadrature components of the potential or current responses to a sinusoidal perturbation is employed to characterize the system impedance. Bridge methods are capable of high accuracy but are rather time-consuming and laborious. In phase-sensitive detection methods, separation of phase shifts associated with the cell impedance from those attributable to the potentiostat sometimes can be troublesome.[3] Gain-phase measurements perhaps offer the most convenience and flexibility in this regard. While an early criticism of this approach was that measurements have to be performed at each frequency in a serial fashion, the advent of modern automated equipment for such measurements has largely overcome this handicap.

3. *Treatment of Data*

The basis for most impedance or admittance measurements lies in an effective means of relating the measured *frequency-dependent* parameters to *passive* network elements comprising the specific test system. Admittance representations (Equation 7) are useful for parallel components while circuit elements connected in series are more commonly described in terms of an impedance (Equation 6). However, as we show below, conversion from a parallel to series representation or *vice versa* frequently facilitates data interpretation and resolution of individual network components. It is important to re-emphasize that both admittance and impedance as measured are complex quantities comprising conductance (real) and susceptance (imaginary) terms in the former, and resistance (real) and re-

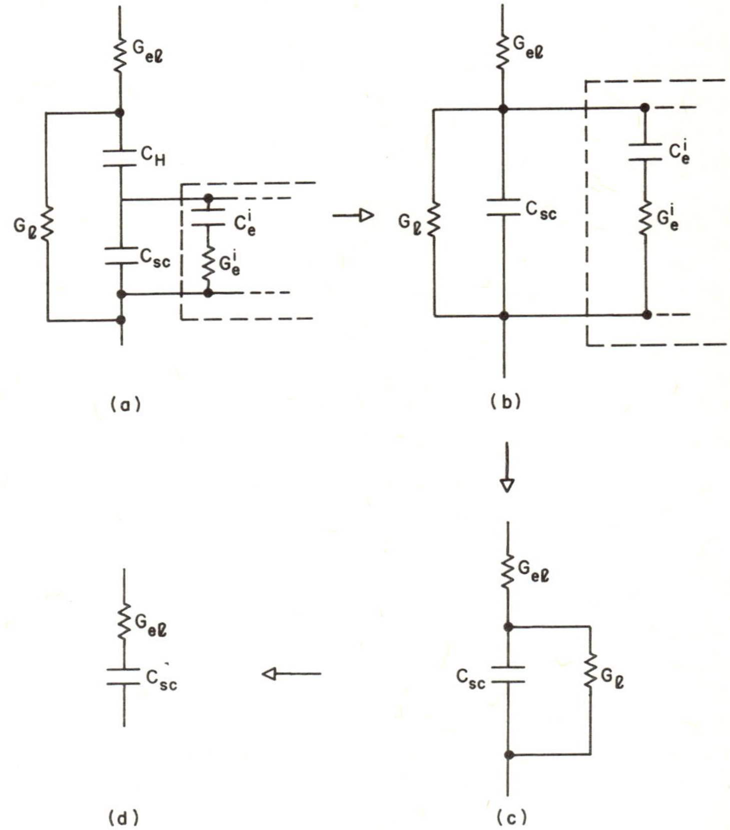

FIGURE 1. Successive reductions in the equivalent circuit for a semiconductor/molten salt electrolyte interface. A description of symbols is given in the text.

actance (imaginary) terms in the latter (Section 2). Furthermore, representation of the test system in terms of passive elements may not always be appropriate. In situations where there is charge transfer across the interface and mass transfer effects are important, introduction of reactive elements (e.g., Warburg impedance) becomes necessary.[3,5] For the present application at least, such complications may be avoided by appropriate choice of experimental conditions such that the interface may be treated as ideally polarizable or nearly so. Comparison of computer simulation models with the behavior actually observed for "real" systems shows that this is indeed a valid assumption for selected systems *(vide infra)*.

A second question is whether the measured impedance (or admittance) truly represents only the semiconductor/electrolyte interface (which is the desired situation) or some ill-defined portion of the composite electrochemical cell as a whole. Again by judicious selection of cell geometry, the interface can be isolated from the residual cell components. For example, the counterelectrode area can be made orders-of-magnitude larger than the semiconductor working electrode so that only the capacitance of the latter is measured. The electrode spacing can be adjusted depending on the conductivity of the electrolytes. (This is especially important in systems employing low-temperature molten salt electrolytes in view of the fact that they have significantly lower conductivity than aqueous supporting electrolytes).

Finally, the measured parameters can be displayed either in terms of plots of the in-phase and quadrature components separately as functions of frequency or together as a vector in the complex plane. The latter procedure is analogous to Nyquist diagrams or Cole-Cole plots and has been employed for the representation of electrode kinetics.[2] For the present application, however, the former procedure seems more appropriate and therefore will be adopted.

4. *Equivalent Circuits for the Semiconductor/Molten Salt Electrolyte Interface*

A general treatment of the semiconductor/electrolyte interface is beyond the scope of this chapter. The reader is referred to reviews in the literature for equivalent circuit representations of the semiconductor/electrolyte interface.[6] Only information germane to the molten salt electrolyte case and to the automated measurement technique to be presented below is detailed herein.

Figure 1 illustrates successive reductions of the equivalent circuit for a semiconductor/molten salt electrolyte interface. Relatively little is known on the nature of the electrical double layer in molten salts at metallic electrodes. Earlier work in this field has been comprehensively reviewed and much of the research activity has been by Russian investigators.[7a–f] Differential capacitances have been measured at a hanging mercury drop electrode as a function of melt acidity in the $AlCl_3$-1-butylpyridinium chloride (BPC) system.[7g] Knowledge of the electrical double layer at semiconductor electrodes is scant too, although progress has been made in this direction on oxide semiconductors in contact with high-temperature chloroaluminate melts.[8] As in the case of concentrated aqueous electrolytes, the Gouy layer capacitance has been omitted in the network representation shown in Figure 1a for a semiconductor/molten salt interface. In series with this capacitive term is the Helmholtz layer capacitance, C_H. C_{sc} is the capacitance of the space-charge layer in the semiconductor (this layer, unlike that in the case of metal electrodes, penetrates significantly into the semiconductor because of the much lower free-carrier density in the latter case). The width of this layer can be varied by varying the applied potential.[6] This capacitance is in series with C_H and a composite conductance element, G_{el}, comprising the contributions from the electrolyte, space-charge layer, semiconductor bulk, and the rear ohmic contact. The conductance term G_l ($G_l = 1/R_{sh}$, R_{sh} = shunt resistance) describes the extent of leakage (or polarizability) of the semiconductor/molten salt electrolyte contact. The series elements, G_{el} and G_e in Figure 1 are more appropriately described by resistances, R_{el} and R_e, respectively. However, the conductance representation, $G = 1/R$, is retained since G is the measured quantity here. The "excess" series R-C elements denoted by the subscripts "e" are connected in parallel with C_{sc} and represent a variety of charge accumulation modes in the interface. Charge accumulation at the interface has been represented in terms of *frequency-dependent* R-C elements in parallel with C_{sc} in recent studies on semiconductor/ aqueous electrolyte interfaces.[9] The usefulness of such equivalent circuit representations in terms of understanding the physico-chemical nature of interfacial processes, however, seems questionable. Charge accumulation can arise from deep or shallow interface states, either intrinsic or extrinsic to the electrode. Such states, which give rise to trapping of electronic carriers and therefore to accumulation of charge, are caused by defects in the solid semiconductor, presence of dopant species, or adsorption of species from the electrolyte.[10] Either a limited number of *discrete* R-C elements each exhibiting a distinct relaxation time or a large number of overlapping elements with a *continuous* distribution of relaxation times can ensue depending on the specific physical and chemical state of the test interface.[11]

The network as shown in Figure 1a is completely general in that it can describe any given semiconductor/molten salt electrolyte interface provided that the interface is polarized to a fairly high degree. The presence of significant faradaic charge transfer across the interface would necessitate the inclusion of additional circuit elements in *series* with C_{sc}, perhaps an equivalent of the Randles circuit.[12] Such a situation, however, as stated before, is not germane to the experimental cases to be described below.

Since C_H is connected in series with C_{sc} and the assumption can be made that $C_H \gg C_{sc}$, the circuit in Figure 1a reduces to that in Figure 1b. The excess R-C terms alternatively can be removed by the procedure described in Section 5.3 below to yield the three-element circuit shown in Figure 1c. Finally the shunt term, G_l can be determined (cf. Section 5.2, *vide infra*) so that a single series G_{el}–C_{sc} term (Figure 1d) remains. From the measured relaxation of this R–C element, C_{sc} and G_{el} are determined for the given interface as detailed in the next section.

5. *Methodology*

5.1. *Determination of* C_{sc} *and* G_{el}

The admittance, Y_d, of the circuit shown in Figure 1d is given by Equation 8:

$$Y_d = \frac{\omega^2 G_{el}C_{sc}^2 + j\omega G_{el}^2 C_{sc}}{G_{el}^2 + (\omega C_{sc})^2} \tag{8}$$

The measured admittance, Y_m, is given by Equation 9:

$$Y_m = G_p + j\omega C_p \tag{9}$$

Here the term, C_p, although shown as a parallel capacitive element, may also be equally well represented as a susceptance term (see Section 2). Equating Equations 8 and 9 and rearrangement yields Equations 10 and 11 relating the measured parameters with passive equivalent circuit elements.

$$G_p/\omega = \frac{\omega C_{sc}^2 G_{el}}{G_{el}^2 + (\omega C_{sc})^2} \tag{10}$$

$$\omega C_p = \frac{\omega G_{el}^2 C_{sc}}{G_{el}^2 + (\omega C_{sc})^2} \tag{11}$$

According to Equations 10 and 11, plots of ωC_p *vs.* ω and G_p/ω *vs.* ω will have maxima at a frequency, $\omega_{max} = G_{el}/C_{sc}$. The reciprocal of ω_{max} yields of course the relaxation time, τ, of the G_{el}–C_{sc} element. The functions G_p/ω and ωC_p in Equations 10 and 11 will have the values at ω_{max} given by Equations 12 and 13:

$$G_p/\omega_{\,max} = C_{sc}/2 \tag{12}$$

$$\omega C_{p\,max} = G_{el}/2 \tag{13}$$

This readily follows by differentiating Equations 10 and 11 with respect to ω and setting the product to zero.

5.2. *Determination of* G_l

Consider now the addition of G_l to the circuit shown in Figure 1d (c.f. Figure 1c). The admittance of the circuit in Figure 1c is given by Equation 14:

$$Y_c = \frac{G_{el}G_l(G_{el} + G_l) + \omega^2 G_{el}C_{sc}^2 + j[\omega G_{el}C_{sc}(G_{el} + G_l) - \omega C_{sc}G_{el}G_l]}{(G_{el} + G_l)^2 + \omega^2 C_{sc}^2} \tag{14}$$

Again, this equation may be equated with that describing the measured admittance, Equation 9, to yield Equations 15 and 16:

$$\omega C_p = \frac{\omega G_{el}C_{sc}(G_{el} + G_l) - \omega C_{sc}G_{el}G_l}{(G_{el} + G_l)^2 + (\omega C_{sc})^2} \tag{15}$$

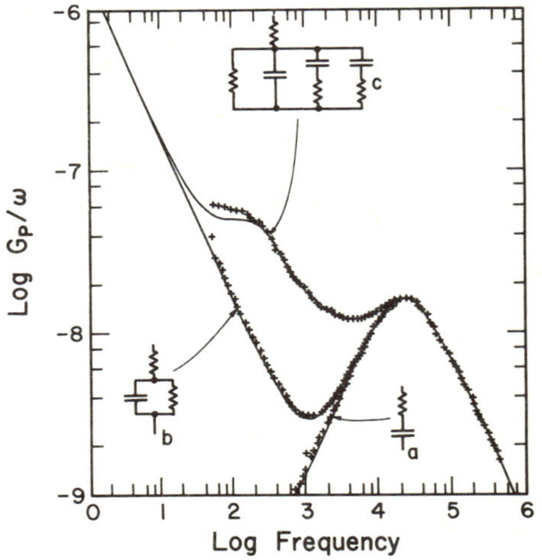

FIGURE 2. Experimental (crosses) and predicted response (solid lines) of model circuits in terms of G_p/ω *vs.* ω plots. The lines labelled 'a', 'b', and 'c' represent, respectively, the behavior predicted by the circuits in Figures 1d, 1c, and 1b. For the circuits in Figure 1b, the special case when $i = 2$ is considered here (see text). For computation of circuit response, the following parameters were assumed: $C_{sc} = 32{,}900$ pF; $G_{el} = 5 \times 10^{-3}$ mho; $G_l = 1 \times 10^{-5}$ mho; $G_e^1 = 8.2 \times 10^{-5}$ mho; $G_e^2 = 1.2 \times 10^{-4}$ mho; $C_e^1 = 75{,}900$ pF, and $C_e^2 = 9000$ pF.

$$G_p/\omega = \frac{(G_{el}G_l/\omega)(G_{el} + G_l) - \omega G_{el}C_{sc}G_l}{(G_{el} + G_l)^2 + (\omega C_{sc})^2} \tag{16}$$

Figure 2 illustrates the G_p/ω *vs.* ω response of the circuits shown in Figures 1d (curve a) and 1c (curve b), respectively. The crosses represent experimental data on model circuits using the automated technique described below. The solid lines are the responses predicted by Equations 10 and 16, respectively. Precision low-loss capacitors and metal film resistors were used to assemble the model circuits. The values chosen for these circuit elements are typical of those observed for semiconductor/molten salt electrolyte junctions (see Section 9).

The point to note in Figure 2 is that inclusion of G_{el} in the two-element circuit in Figure 1d affects predominantly the low-frequency regime. To illustrate how the admittance at very low frequencies ($\omega \rightarrow$ dc) reduces to G_l, Equation 16 is multiplied by ω and evaluated at $\omega = 0$, i.e.,

$$\left. G_p \right|_{\omega = 0} = \frac{(G_{el}G_l)(G_{el} + G_l)}{(G_{el} + G_l)^2} = \frac{G_{el}G_l}{(G_{el} + G_l)} \tag{17}$$

Since G_{el} is usually orders of magnitude larger than G_l (see below), Equation 17 reduces to Equation 18:

$$\left. G_p \right|_{\omega = 0} \cong G_l \tag{18}$$

5.3. *Determination of* G_e^i *and* C_e^i

The complex admittance of the circuit shown in Figure 1b is given by Equation 19:

$$Y = G_{el} \parallel \left(j\omega C_{sc} + G_l + \sum_i Y_e \right) \tag{19}$$

where

$$Y_e = \frac{\omega^2 C_e^2 G_e + j\omega C_e G_e^2}{G_e + (\omega C_e)^2} \tag{20}$$

In Equations 19 and 20, G_e and C_e are the conductance and capacitance associated with each charge accumulation mode at the interface (cf. Figure 1b). As in the above cases, the total admittance in Equation 19 may be resolved into its real and imaginary parts by equating with the corresponding terms of the measured admittance (cf. Equation 9). The procedure we adopt for determining G_e and C_e is best illustrated by considering only two charge accumulation modes such that two series elements, G_e^1-C_e^1 and G_e^2-C_e^2, are connected in parallel with C_{sc} (cf. Figure 1b). These two charge accumulation modes could either comprise a *single* process with two distinct relaxation times (e.g., surface-state charging) or two different processes at the interface.

The G_p/ω *vs.* ω response of such a model circuit is illustrated in Figure 2. The difference between curves 'b' and 'c' represents the contribution of charge accumulation modes to the total admittance of the interface. Again, the crosses in Figure 2 are the experimental data and the solid line is the response predicted by Equation 19 for $i = 2$. More detailed analyses are presented below.

The presence of G_{el} in series with the other elements in Figure 1b entails the sequence of transformations illustrated in Figure 3. The measured (frequency-dependent) conductance, G_m, and the capacitance, C_m, terms are converted to their series equivalents (Figure 3b) from which $R_{el} = 1G_{el}$ is subtracted and reconverted back to parallel elements to yield the circuit in Figure 3c. G_{el} is determined by the procedure outlined in Section 5.1. This circuit is equivalent to the representation shown in Figure 3d in terms of passive elements. The series G_e–C_e terms for the two modes are converted to their parallel equivalents (cf. Figure 3c). At this stage, all conductances and capacitances are *in parallel* facilitating the removal of G_l and C_{sc} (determined by procedures outlined in Sections 5.2 and 5.1, respectively), to yield the circuit in Figure 3f. Now the final resolution into the individual elements for the two assumed charge accumulation modes remains to be accomplished.

Consider the G_p/ω *vs.* ω (Figure 4a) and ωC_p *vs.* ω (Figure 4b) responses of the circuit in Figure 1b for the present case when $i = 2$. The curves labelled

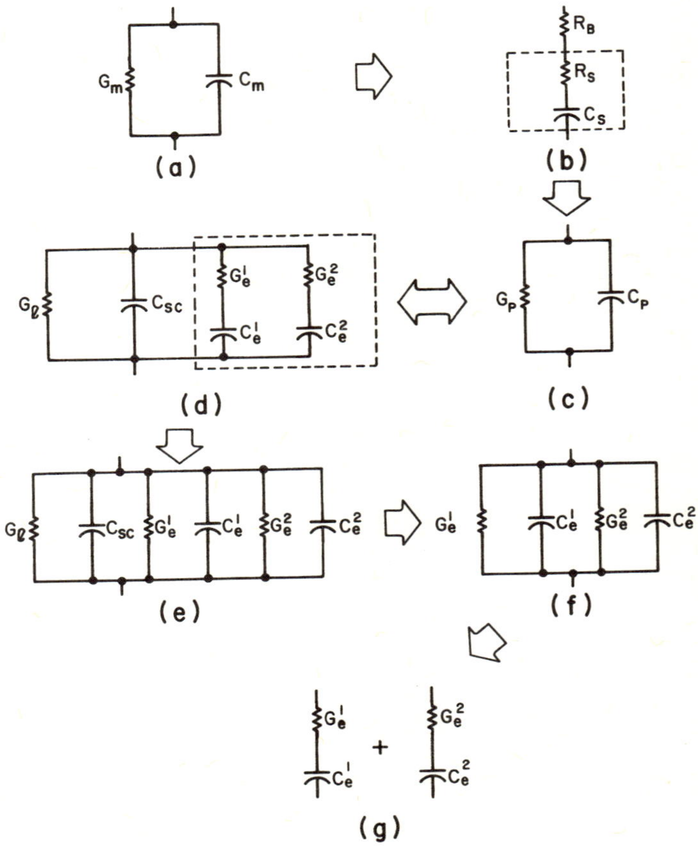

FIGURE 3. Sequence of transformations for extracting G_e^1, G_e^2, C_e^1, and C_e^2 from the measured admittance of the semiconductor/molten salt electrolyte interface.

'a–e' correspond to the circuits in Figure 5. Two alternative approaches may be adopted to resolve the composite curves (labelled 'c') in Figure 4a and 4b. In the first, the dominant peak in ωC_p *vs*. ω (Figure 4b, curve 'c') is assumed to correspond to one charge accumulation mode. The parameters for this mode are extracted from Equations 21 and 22:

$$\omega C_p|_{max} = G_e^1/2 \tag{21}$$

$$\text{and } 1/\omega_{max}^1 = C_e^1/G_e^1 \tag{22}$$

to yield the curve labelled 'e' in Figure 4 for the G_e^1-C_e^1 element. The parameters for the other mode (G_e^2, C_e^2) may then be computed from Equation 23 and Equation 24[11]:

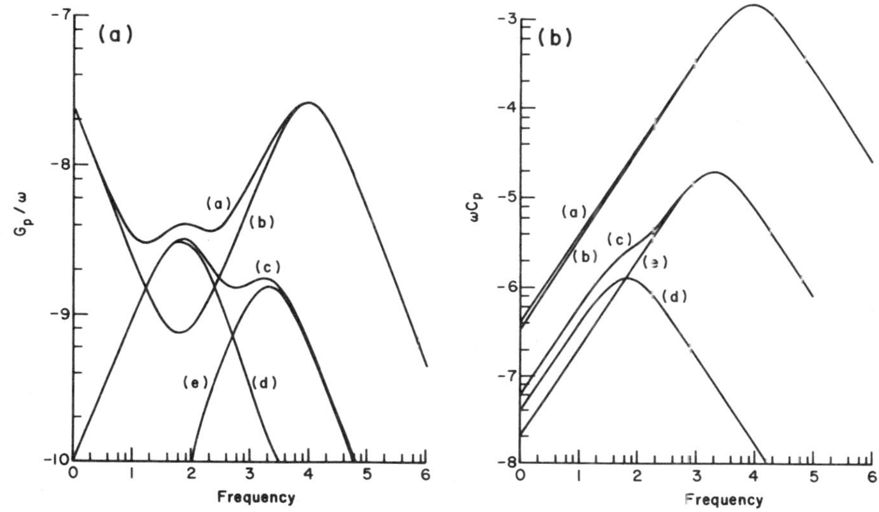

FIGURE 4. G_p/ω *vs.* ω (Figure 4a) and ωC_p *vs.* ω (Figure 4b) response of a model circuit (cf. Figure 1b with i = 2). The labels 'a–e' denote the response for versions of the circuit shown in Figure 5.

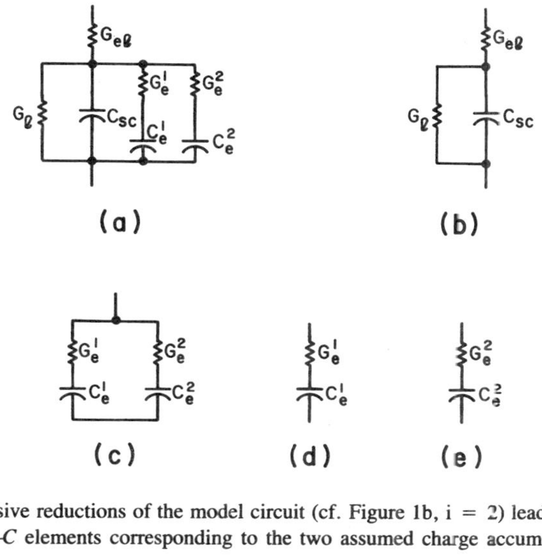

FIGURE 5. Successive reductions of the model circuit (cf. Figure 1b, i = 2) leading finally to the individual series *G–C* elements corresponding to the two assumed charge accumulation modes in the system.

$$G_p/\omega|_{max} = C_e^2/2 \quad (23)$$

$$1/\omega_{max}^2 = C_e^2/G_e^2 \quad (24)$$

The G_e^2–C_e^2 response yields the curves denoted by 'd' in Figure 4.

Alternatively, the dominant peak of curve 'c' in Figure 4a is utilized and G_e^2 and G_e^2 are determined using Equations 23 and 24, respectively, to yield the response shown as curve 'd'. The parameters for the other charge accumulation mode may be obtained then by simple subtraction. Because of the discrepancy induced by overlapping time constants (compare curves 'c' and 'd' in Figure 4), the new parameters determined thus from curve 'e' may be used to redetermine G_e^2 and C_e^2. This iterative procedure is continued until convergence is obtained with the experimental data.

The advantage of the first approach is simply that the relative insensitivity of C_p to charge accumulation modes (see below) is exploited obviating the need for iteration (e.g., compare curves 'c' and 'e' in Figure 4b).

The methods outlined above were again tested on a model circuit. The individual component values were first determined on a General Radio Model 1616 ac bridge. These values are assembled in Table I. Also shown for comparison are the values determined by the present method and the percent deviation from the values measured on the bridge. The error is seen to be $\leqslant \sim 2\%$ for all the components except G_e^2 and C_e^2. The somewhat larger deviations in this case may be attributed to the proximity of the G_e^2–C_e^2 relaxation to the peak caused by the G_{el}–C_{sc} relaxation (cf. Figure 6a). The circuit components listed in Table 1 were selected such that they provided a good representation of the behavior of an actual semiconductor/molten salt electrolyte interface (see Section 9). The overlapping time constants in the model circuit and the attendant errors in Table 1, therefore, should be regarded as a "worst possible case."

TABLE I. Comparison of the Component Values for the Model Circuit (cf. Figure 5a) Measured on an AC Bridge with Values Determined by Automated Admittance Spectroscopy

	Measured value		
Component	Bridge	Admittance technique	% Error
C_{sc}, pF	54,300	53,000	2.4
G_l, mho	1.44×10^{-7}	1.44×10^{-7}	0.0
G_{el}, mho	3.01×10^{-3}	2.98×10^{-3}	1.1
G_e^1, pF	6,260	6,280	0.3
G_e^1, mho	2.54×10^{-6}	2.50×10^{-6}	1.6
C_e^2, pF	3,132	2,752	12.1
G_e^2, pF	3.92×10^{-5}	4.18×10^{-5}	6.6

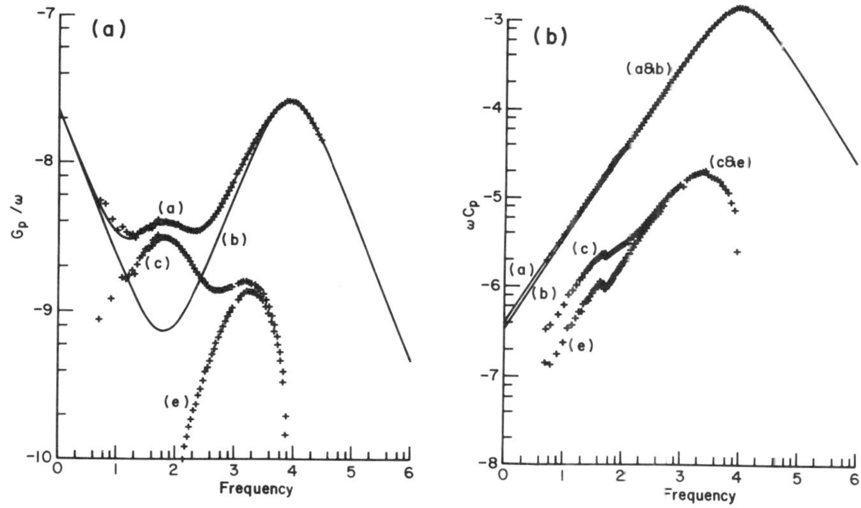

FIGURE 6. Experimental (crosses) and predicted response (solid lines) of the model circuits in Figure 5. The labels 'a–e' correspond to the circuits in Figure 5. Parametric values assumed for generating the response curves are assembled in Table 1. Data are shown as G_p/ω *vs.* ω response (Figure 6a) and ωC_p *vs.* ω plots (Figure 6b).

The G_p/ω *vs.* ω and ωC_p *vs.* ω plots for the model circuit (cf. Figure 1a, $i = 2$) are illustrated in Figures 6a and 6b, respectively. The crosses again are the experimental data and the solid curves denote the behavior predicted by the model described above. The curve labelled 'b' is the response of the three-element G_{el}–G_l–C_{sc} circuit (cf. Figure 1b). The set of data marked 'c' represents the composite response deduced from the circuit transformations shown in Figure 3. Those labelled 'e' denote the relaxation of the G_e^2–C_e^2 term. The relaxation of the series G_e^1–C_e^1 combination is omitted from these figures for the sake of clarity.

6. *Computer Simulation of Model Circuits*

Computer simulations permit examination of the sensitivity of the circuit response (in terms of G_p/ω *vs.* ω plots, for example) to individual component values and also delineation of the frequency range(s) where the influence of each circuit element or combinations thereof is expected to be predominant. This is done in the usual manner by holding all the other terms constant while varying the magnitude of the circuit element(s) of interest.

Figure 7 illustrates the response of the three-element circuit (cf. Figure 1c) wherein G_l is varied while maintaining G_{el} and C_{sc} constant. The circuit response

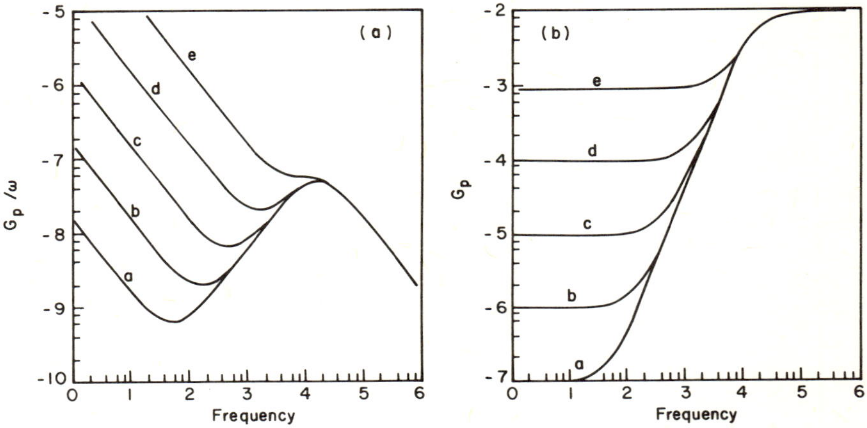

FIGURE 7. Computer simulation of G_p/ω *vs*. ω response (Figure 7a) and G_p *vs*. ω response (Figure 7b) for the three-element circuit model (cf. Figure 1c). The parameters were the following: C_{sc} = 0.1 μF; G_{el} = 10^{-2} mho; and G_F = (a) 10^{-7} mho, (b) 10^{-6} mho, (c) 10^{-5} mho, (d) 10^{-4} mho, and (e) 10^{-3} mho.

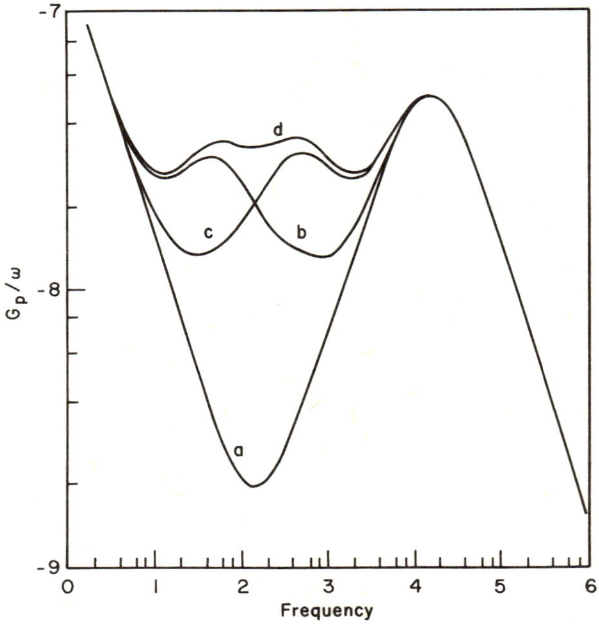

FIGURE 8. Computer simulation of G_p/ω *vs*. ω response for the model circuit (cf. Figure 5a) with parameters whose values are the following: C_{sc} = 0.1 μF; G_{el} = 10^{-2} mho, G_l = 10^{-6} mho; $\omega^1_{e(max)}$ = 50 Hz; $\omega^2_{e(max)}$ = 500 Hz; and (a) no charge accumulation; (b) C^1_e = 0.05 μF; C^2_e = 0.01 μF; (c) C^1_e = 0.01 μF, C^2_e = 0.05 μF; and (d) C^1_e, C^2_e = 0.05 μF.

is shown in terms of G_p/ω *vs.* ω (Figure 7a) and G_p *vs.* ω (Figure 6b) plots. The larger the G_l term, the higher the frequency at which $G_p \rightarrow G_l$ (cf. Figure 6b). The smaller the G_l term, the more clearly defined is the relaxation of the G_{el}–C_{sc} term (cf. Figure 7a). The importance of ensuring a high degree of polarization of the interface in determinations of C_{sc} is therefore quite obvious.

Figure 8 illustrates three examples of computer simulations of the circuit in Figure 1b (for the specific case where $i = 2$) wherein C_{sc}, G_{el}, G_l, and the peak frequencies corresponding to G_e^1–C_e^1 and G_e^2–C_e^2 terms are held constant while C_e^1 and C_e^2 are systematically varied. The degree of separability of the individual G_e and C_e elements (cf. Section 5.3) will be critically dependent on their *RC* time constants and in the limiting case, a continuum of levels will be obtained. As discussed by previous authors,[11] this dispersive effect will be reflected in a broadening of the G_p/ω *vs.* ω response curves.

What influence do the charge accumulation modes have on the relaxation of the G_{el}–C_{sc} term? Again the answer is to be found in the relative magnitude of the time constants corresponding to the two processes. For magnitudes typical of G_{el} and C_{sc} for semiconductor/electrolyte interfaces, the relaxation frequency of the G_{el}–C_{sc} combination occurs at fairly high frequencies. Therefore, the charge accumulation mode will have to be fairly fast (e.g., fast surface states) in order that its time constant overlaps with the G_{el}–C_{sc} relaxation process. On the other hand, in the presence of very slow states, resolution of the individual G_e, C_e terms from G_l will be hindered by overlap of the time constants corresponding to the respective relaxations.

7. *Application to Semiconductor/Electrolyte Interfaces*

Two major applications of the methodology for the study of semiconductor/electrolyte interfaces, presented in preceding sections, is for determinations of the semiconductor flat-band potential, V_{fb}, and for the characterization of "low-frequency" effects (mainly estimation of G_l and characterization of charge-accumulation modes). Each of these application areas is explored below.

7.1. *Determination of V_{fb}*

The flat-band potential is a fundamental parameter in semiconductor electrochemistry.[13] Knowledge of this parameter enables the mapping of semiconductor energy levels at the electrode/electrolyte interface. Such a mapping plays a key role in the evaluation of a given electrode/redox electrolyte combination for photoelectrochemical applications, since the theoretical maximum output from this system is often determined by the difference between V_{fb} and the redox potential.[14] The V_{fb} values are usually obtained from measurements of the differential capacitance C, as a function of applied potential V, of an electrochemical cell comprising a semiconductor electrode and a counterelectrode. An *n*-type semiconductor is assumed for this discussion. Under conditions such that a depletion region exists on the semiconductor surface, the Mott-Schottky relation is obeyed and a plot of $1/C^2$ *vs.* V yields a straight line whereby V_{fb} and the semiconductor doping density, N_D, can be determined from the intercept and the slope, respectively, from Equation 25[6c]:

$$1/C^2 = \frac{2}{q\varepsilon_0\varepsilon_s N_D A^2}(V - V_{fb} - kT/q) \tag{25}$$

In Equation 25, ε_0 = permittivity of free space, ε_s = semiconductor permittivity, A = electrode area, T = temperature, and the other terms have their usual significance.

In the above analysis, the tacit assumption is often involved that C refers solely to the space-charge-layer capacitance in the semiconductor, although it has been shown[15] that the Mott-Schottky relation is not invalidated *per se* even in cases where there are nontrivial contributions to the measured capacitance from effects in the Helmholtz region (e.g., specific ion adsorption, acid-base equilibria). In such cases, the Mott-Schottky plots show a parallel shift from the behavior predicted by Equation 25, the intercept potential, V_0, now given by Equation 26[15,16]:

$$V_0 = V_{fb} + kT/q - q\varepsilon_0\varepsilon_s N_D A^2/2 C_H^2 \tag{26}$$

Another complication in the interpretation of Mott-Schottky data arises from frequency dispersion of the measured capacitance values. In such cases, as discussed by previous authors,[17–20] unique assignment of the measured slopes and intercepts becomes impossible. As we shall attempt to illustrate below, the present methodology enables the extraction of frequency-independent Mott-Schottky parameters. Additionally, *a priori* characterization of the various components of the electrical network comprising the semiconductor/electrolyte interface (cf. Figure 1) locates frequency regimes which are unclouded by dispersion phenomena and locates an optimum range for subsequent measurement.

7.2. *Low-Frequency Effects*

To appreciate the importance of determining G_l for a given semiconductor/electrolyte interface for solar photovoltaic applications, consider the equivalent circuit shown in Figure 9 for an operating solar cell.[21,22] In this representation, the cell is shown as a constant current generator shunted by G_l. Obviously, the smaller the G_l term, the larger will be the power available at the output terminals of the cell to an external load. Comparison of G_l for a variety of systems, estimated by the method outlined in Section 5.2, will yield a diagnostic criterion

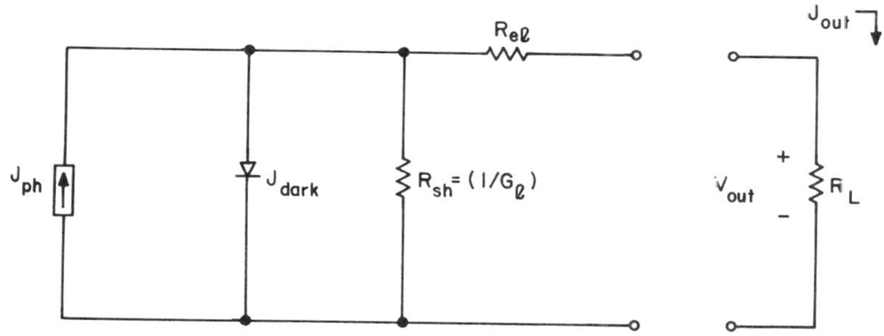

FIGURE 9. Equivalent circuit of an operating solar cell. J_{ph} = photocurrent; J_{dark} = dark current; $R_{e\ell}$ = series resistance; J_{out}, V_{out} = output current and voltage; and R_L = load resistance.

for the efficacy of a given semiconductor/electrolyte combination for photovoltaic applications.

Charge accumulation modes at a semiconductor/electrolyte interface can have important consequences for photovoltaic applications, given that, in many cases, charge transfer across the interface is mediated by band-gap states.[23] The rate at which these "trap" states transfer charge to appropriate redox species in the electrolyte will determine the extent of deleterious carrier recombination at the surface.[23] The influence of surface treatment of the semiconductor in "tuning"

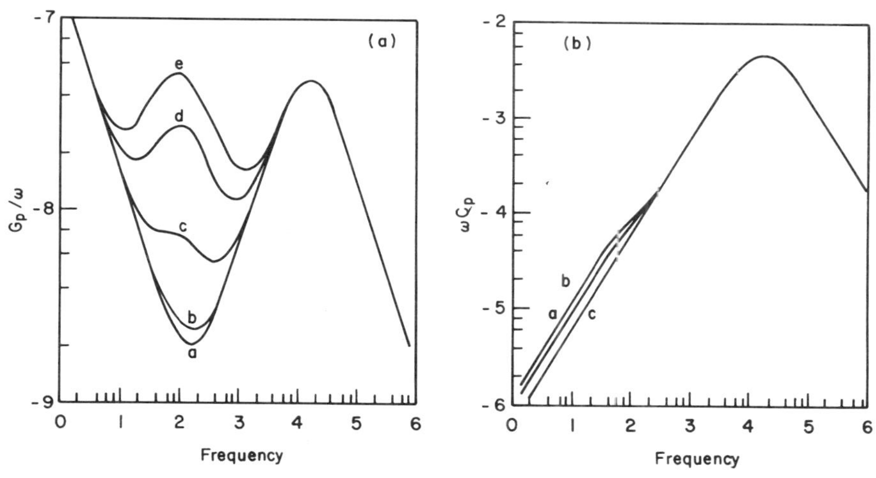

FIGURE 10. Comparison of the relative sensitivity of G_p (Figure 10a) and C_p (Figure 10b) to interfacial processes. Data are displayed in the form of G_p/ω *vs.* ω and ωC_p *vs.* ω plots, respectively. Parameters for computer simulation are as follows: C_{sc} = 0.1 μF; G_{el} = 10^{-2} mho; G_l = 10^{-6} mho. Curve 'a' is the response in the absence of charge accumulation and curves 'b–e' denote response for a circuit with a single charge accumulation mode (cf. Figure 1b, i = 1) with C_e = (b) 0.001 μF, (c) 0.01 μF, (d) 0.05 μF, and (e) 0.1 μF.

the interface to optimal levels of charge transfer may be conveniently monitored in terms of a correlation between the conversion efficiency of the cell and any variations in the relaxation behavior of the G_e–C_e terms in the corresponding equivalent circuit of the interface (cf. Figure 1a).

The extreme sensitivity of conductance measurements to charge accumulation at the interface is illustrated by Figure 10 which shows the results of computer simulations of a model semiconductor/electrolyte interface. In these simulations for a single charge accumulation mode (cf. Figure 1a, $i = 1$), C_e was varied while holding C_{sc}, G_{el}, and G_l constant. The G_p/ω *vs.* ω response (Figure 10a) is far more sensitive to interfacial charge accumulation than the ωC_p *vs.* ω response (Figure 10b). The reason for this is simply that the peak height in ωC_p *vs.* ω curves is equal to $G_{el}/2$ (cf. Equation 13) and since usually $G_{el} \gg G_e$, the effect of the latter will be smeared out.

The high sensitivity of conductance is also the underlying basis for the earlier popularity of surface conductance measurements[6b] for the characterization of semiconductor/electrolyte interfaces.

8. *Experimental Details*

8.1. *Measurement Setup and Procedure*

A schematic of the measurement setup that can be employed for automated admittance spectroscopy is shown in Figure 11. A small amplitude ac signal (5–10 mV p/p) is applied to the test electrode in the cell and to the reference channel of a Hewlett-Packard (HP) network analyzer (Model, 3570 A). This signal is generated on a frequency synthesizer (HP Model 3320 B) which allows automatic frequency-scanning over a range from 50 Hz to 13 MHz at preselected increments. For measurements in the 1 Hz–50 Hz range, a Model 3575A HP network analyzer was employed in conjunction with an oscillator circuit. These measurements were carried out manually. The counterelectrode of the cell is maintained at virtual ground by a high-speed I-V converter (c.f., Figure 11). The output of the I-V converter is connected to the "cell" input channel of the network analyzer as shown in Figure 11.

The synthesizer and network analyzer are controlled by a desk-top calculator (HP Model 9825 A) through an HPIB interface bus. The calculator controls the output signal level and frequency of the synthesizer and also sets the input levels and band width settings of the network analyzer. To enhance the signal-to-noise ratio at each frequency, the calculator instructs the network analyzer to measure the attenuation and phase shift of the ac signal through the I-V circuit a pre-specified number of times before incrementing to the next frequency. Typical frequency scan times are approximately two minutes for 100 frequencies—the scan time being largely determined by the longer signal "transit" times at low frequencies.

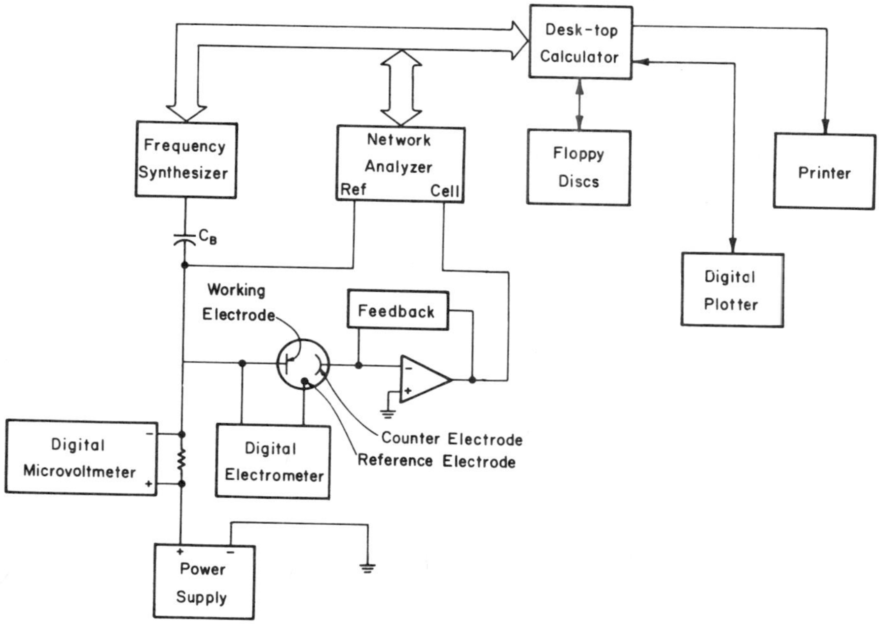

FIGURE 11. Schematic of measurement setup for automated admittance spectroscopy.

All the information on the admittance characteristics of the cell, more specifically the test electrode/electrolyte interface, is contained in the attenuation and phase shift of the ac signal measured at each frequency and bias voltage by the network analyzer. This information is stored on a floppy disc with further data processing being facilitated by the desk-top calculator. These data can then be plotted on a digital recorder or graphics terminal as a function of selected parameters or displayed as a computer readout on a line-printer.

The I-V converter referred to in the preceding paragraphs plays a key role in these measurements in that it equalizes the input independence in the two channels as seen by the network analyzer (cf. Figure 11). Further details on the role of this operational amplifier circuit may be found elsewhere.[24,25] A Teledyne Philbrick Model 1435 operational amplifier was found to have a gain-bandwidth product large enough to ensure no "roll-off" for frequencies from 50 Hz to ~0.1 MHz. The upper limit of the frequency range may be extended even further by compensation circuits used in conjunction with the I-V converter. For the present application, however, the high-frequency limit of ~0.1 MHz was found to be adequate *(vide infra)*.

Consider the I-V converter circuit shown in Figure 12. This circuit is equivalent to that shown in Figure 11 except that the electrochemical cell is now modelled as a parallel *R-C* combination. It is important to emphasize that this cell representation is valid strictly only for the conditions specified in Section

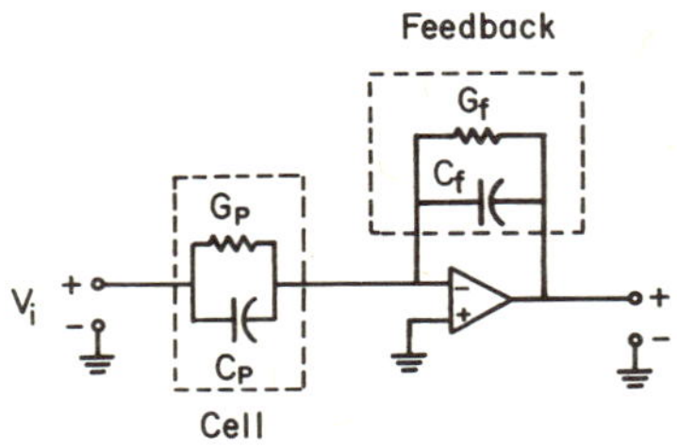

FIGURE 12. I-V converter circuit for automated admittance spectroscopy. V_i and V_o are the input and output signals, respectively. Other symbols are described in the text.

4. The importance of choosing experimental conditions such that the stated criteria are applicable is therefore quite obvious. Assuming infinite amplifier gain, the transfer function $T(j\omega)$ for the circuit shown in Figure 12 is given by Equation 27:

$$T(j\omega) = \frac{(G_p + j\omega C_p)}{(G_f + j\omega C_f)} \tag{27}$$

where G_f and C_f are the feedback elements in the I-V converter circuit (cf. Figure 12) which are preselected at each bias voltage (see below).

Separation of Equation 27 into the magnitude (attenuation) and phase components of $T(j\omega)$ yields, respectively, Equations 28 and 29:

$$T(j\omega) = \frac{(G_p^2 + \omega^2 C_p^2)^{1/2}}{(G_f^2 + \omega^2 C_p^2)^{1/2}} \tag{28}$$

$$< T(j\omega) = \tan^{-1}\left(\frac{\omega C_p}{G_p}\right) - \tan^{-1}\left(\frac{\omega C_f}{G_f}\right) \tag{29}$$

Since G_f and C_f are known and the quantities $T(j\omega)$ and $< T(j\omega)$ are measured by the network analyzer at each bias voltage and frequency, Equations 28 and 29 can be solved for the two unknowns, G_p and C_p.

The voltage across the test cell can be controlled automatically by the desktop calculator and A/D and D/A converters via either a low-noise power supply or a minipotentiostat. When ohmic resistance between the working and reference electrodes is expected to be large as would usually be the case with molten salt electrolytes, use of the latter coupled with feedback compensation is obviously preferable. For the model measurements to be described below, the potentials were applied manually and the test electrode potential was measured on a high-impedance microvoltmeter after sufficient time elapsed to ensure equilibration of residual faradaic processes at the working electrode/electrolyte interface (~2–3 hr). Ohmic resistance between the working and reference electrodes was accounted for by positive feedback compensation and by the use of a Luggin probe (not shown in Figure 11).

To complete the measurement setup, a thermostat may be incorporated into the cell assembly, especially when precise temperature control is desired. Experiments may even be designed with temperature as an additional variable parameter, especially when the contribution of deep trapping levels to the interfacial impedance is expected to be important.

8.2. Calibration

The I-V converter works best when the feedback components match G_p and C_p (cf. Figure 12). Obviously *a priori* optimal matching on an uncharacterized test system (e.g., a semiconductor/molten salt electrolyte interface) is difficult. Therefore, the extent of sensitivity of G_p and C_p to nonoptimal choice of G_f and C_f was assessed as follows. G_f or C_f was set at a predetermined value and G_p or C_p components both higher and lower than this value were inserted in the circuit in Figure 12. From the measured transfer function of the circuit (cf. Equation 27) and the known G_f (or C_f), G_p (or C_p) could be determined with frequency as a parameter. This value was then compared against that measured on the bridge.

Typical results are illustrated in Figures 13a and 13b for resistance ($1/G_p$) and capacitance, respectively. In these plots the ordinate is the percent error in the measured value from that determined on the bridge. Note that this error is cumulative and comprises the contributions from the error in the bridge measurements of G_p and C_p and from the tolerance limits in the manufacturer-quoted value for G_f and C_f.

From the data in Figure 13, it is obvious that a nominal accuracy in G_p and C_p of ± 5 % is easily attained over the frequency range from 50 Hz to 0.1 MHz, except when very low values of R_p are measured. This problem arises mainly from the constraint in the choice of G_f (or C_f) imposed by the 50-ohm input impedance of the network analyzer. Values of R_f lower than 50 ohm will cause loading of the I-V converter circuit and consequent errors in the measurement of G_p and C_p. Fortunately, low values of R_p for semiconductor/electrolyte interfaces may be avoided by proper choice of experimental conditions. In general, a lower value for R_f of 500 ohm and an upper limit of 50,000 pF for C_f can be set within which measurement errors in G_p and C_p will be within the ± 5 % level over the dynamic range from 50 Hz to 0.1 MHz (cf. Figure 13).

Another problem with measurements on equivalent R-C combinations, discussed in detail elsewhere[24] and not specifically addressed in Figure 13, is that for frequencies well below or above the pole frequency of a given R-C element, the phase angle of the complex admittance tends to 0° and 90° respectively. For these cases, only the larger admittance component will be measured accurately. In general, for a total measurement accuracy in attenuation and phase of 0.1 dB and 0.1°, respectively (well within the capability of the network analyzer used herein), conductance measurements will be accurate to ± 5 % or better from 50

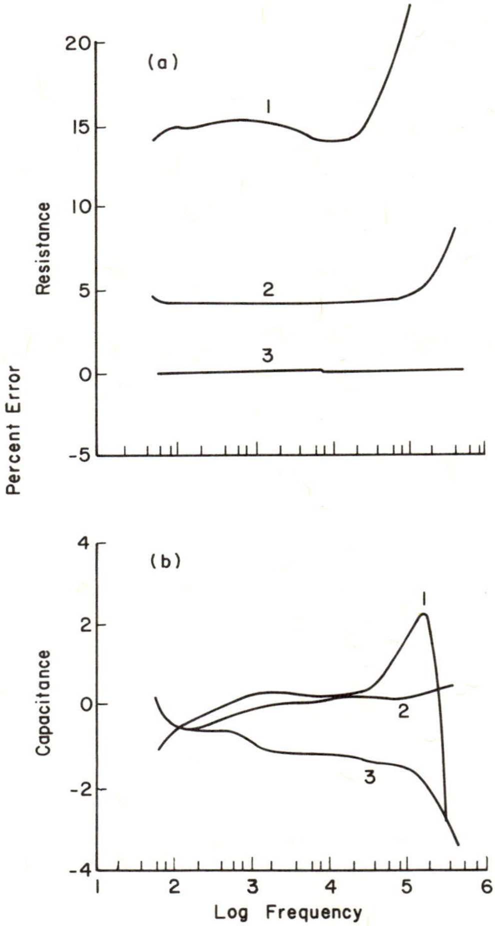

FIGURE 13a. Measurement errors in resistance with a 500-ohm feedback registor; R_p is 10 ohm, 150 ohm, and 500 ohm in 1, 2, and 3, respectively.
FIGURE 13b. Measurement errors in capacitance with a 2000-pF feedback resistor. C_p is 20,000 pF, 2000 pF, and 100 pF in 1, 2, and 3, respectively.

Hz to two decades above the pole frequency, whereas capacitance measurements will be accurate to the same degree from two decades below the pole frequency to the high frequency limit of the system (~0.1 MHz in the percent case).

Further confirmation of the efficacy of the measurement setup and circuitry for automated admittance spectroscopy is provided by measurements on selected model circuits of varying complexity as discussed in Sections 5 and 9.

8.3. Electrochemical Cell and Components

A conventional single compartment three-electrode electrochemical cell was used for all experiments. Vitreous carbon was used as the counterelectrode. Typical dimensions of this electrode were 1.5 × 15 × 30 mm and always such that the geometric surface area of this electrode was orders of magnitude larger than that of the working electrode. An Al wire in a finely fritted glass tube was employed as a quasi-reference electrode. All potentials below are quoted with reference to this electrode. All measurements were performed in the dark and under an Ar atmosphere, the latter because of the air sensitivity of the molten salt electrolyte.

The working electrode was either n-GaAs or n-InP fabricated according to procedures detailed elsewhere.[22,26] Both (100) as well as (111B) surfaces were exposed to the electrolyte. These surfaces were nominally etched to a matte finish prior to a measurement. The etchants for n-GaAs were $H_2SO_4 : H_2O_2 : H_2O$ (3 : 1 : 1) and 6 *M* HCl, the latter being primarily used to remove surface oxide layers. Br_2/MeOH was used for pretreating the n-InP surface.

8.4. Electrolytes

Mixtures of $AlCl_3$ and BPC were used as model molten salt electrolytes. For further details on the preparation of these electrolytes and their properties, the reader is referred to Chapter 3 in Volume 1 of this series.

9. Application to Specific Test Interfaces

9.1. The n-GaAs/$AlCl_3$-BPC Interface

Representative data on this interface at two different bias voltages, namely 0.6 V and 1.2 V, are shown in Figure 14. The G_p/ω *vs.* ω plots adhere to the behavior seen with model circuits (compare Figure 14 and curves 'c' and 'd' in Figure 2) confirming the validity of the equivalent circuit models discussed in Section 4. Computer programs have been developed in this laboratory to automatically determine the maximum values of the G_p/ω and ωC_p functions (cf. Equations 12 and 13). For the particular examples shown in Figure 14, C_{sc} was calculated to be 33,300 pF and 21,400 pF at 0.6 V and 1.2 V, respectively. In a similar fashion, G_{el} was calculated from ωC_p *vs.* ω plots (not shown) to be 5.1×10^{-3} mho. The expected insensitivity of G_{el} to the applied bias was confirmed in these analyses. Using the above values of C_{sc} and G_{el} "theoretical" G_p/ω *vs.* ω plots were constructed; these are shown in Figure 14 as solid lines (curves a and a′). The degree of fit of these lines to the experimental data shown as crosses is further evidence that the observed peak is due to the relaxation of

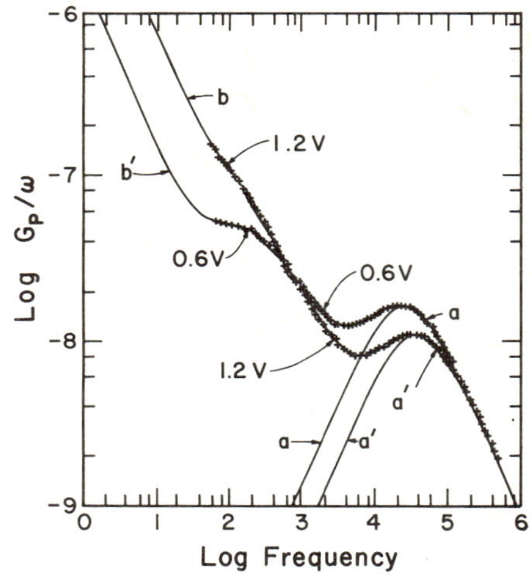

FIGURE 14. Experimental (crosses) and predicted (solid lines) G_p/ω *vs.* ω response of the *n*-GaAs/$AlCl_3$-BPC interface. The $AlCl_3$-BPC mol ratio was ~2 : 1. Data at two bias voltages (0.6 V and 1.2 V) are shown. Labels a,a′ and b,b′ correspond to the circuits in Figure 1d and 1b (for the case when i = 2, refer to text), respectively. Parameters used for generating the response curves were as follows: C_{sc} = 33,300 pF; $G_l = 1 \times 10^{-5}$ mho; $G_e^1 = 8.1 \times 10^{-5}$ mho; C_e^1 = 75,500 pF; $G_e^2 = 1.1 \times 10^{-4}$ mho and C_e^2 = 9000 pF at 0.6 V and C_{sc} = 21,900 pF; $G_l = 5 \times 10^{-5}$ mho; $G_e^1 = 5.8 \times 10^{-5}$ mho; C_e^1 = 65,400 pF; $G_e^2 = 7.5 \times 10^{-5}$ mho; and C_e^2 = 6000 pF. G_{el} was assumed to be 5.1×10^{-3} mho and independent of bias voltage.

the G_{el}–C_{sc} term in the circuit (cf. Section 5.1). The lines shown as b and b′ in Figure 14 represent the G_p/ω *vs.* ω response of the n-GaAs/$AlCl_3$–BPC interface after inclusion of the G_l term (cf. Section 5.2). Note that the G_{el}-C_{sc} relaxation peaks at the two bias voltages are not influenced by the addition of this extra circuit element, substantiating the conclusions reached in Section 5.2.

Using the "frequency-independent" C_{sc} values calculated from the above analysis as a function of bias voltage, a Mott–Schottky plot was constructed for the test interface as shown in Figure 15. A V_{fb} value of 0.180 ± 0.005 V is

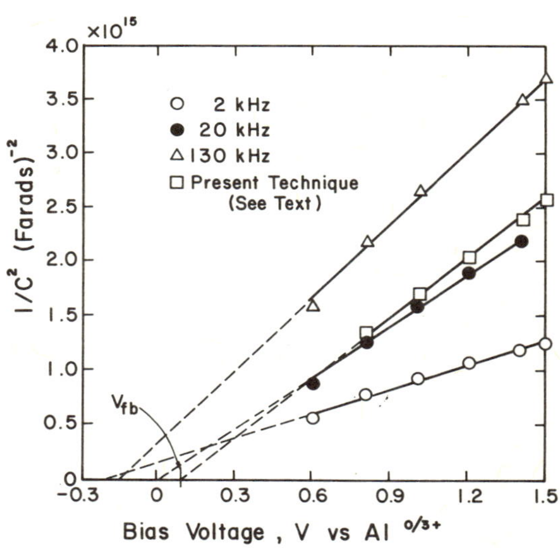

FIGURE 15. Mott-Schottky plots for the *n*-GaAs/$AlCl_3$–BPC interface. The $AlCl_3$–BPC mol ratio was ~2 : 1. Data using both frequency-independent parameters as well as those based on the series capacitance (C_s) values at discrete frequencies (2 kHz, 20 kHz, and 130 kHz) are compared (refer to text).

estimated from this plot. The degree of reproducibility was tested by two separate experiments on the same electrode/electrolyte combination. The calculated C_{sc} values in the two cases agreed to within 2% in the entire bias range. The corresponding V_{fb} values differed by 11 mV. An N_D value of 1.2×10^{17} cm^{-3} was deduced from the slope of the Mott-Schottky plot in Figure 15. For this calculation, ε_s was taken as 10.9 ε_0 ($\varepsilon_0 = 8.85 \times 10^{-14}$ F/cm, cf. Equation 25). This calculated value would be consistent with the Hall value quoted by the manufacturer for this particular electrode (4.5×10^{16} cm^{-3}) if a roughness factor of 1.6 is assumed for the electrode surface area (the geometric surface area was used for A in Equation 25 for the above computation).

As an illustration of the characterization of low-frequency effects (degree of polarizability and charge accumulation), Figure 16 shows typical G_p/ω *vs.* ω (Figure 16a) and ωC_p *vs.* ω (Figure 16b) responses for the model n-GaAs/$AlCl_3$–BPC interface. These data were obtained at a bias voltage of 0.40 V. The plots in Figure 16 are to be compared with the corresponding response of model circuits (cf. Figure 6). As in the latter, crosses denote the experimental data and the solid lines show the response predicted by model (cf. Section 5). As in Figure 6, the solid line labelled 'a' is the response of the complete circuit with two charge accumulation modes (cf. Figure 5a), whereas the line labelled 'b' shows the response of the three-element G_{el}-G_l-C_{sc} circuit (cf. Figure 5b). Obviously, the behavior of the n-GaAs/$AlCl_3$–BPC is defined completely only when two

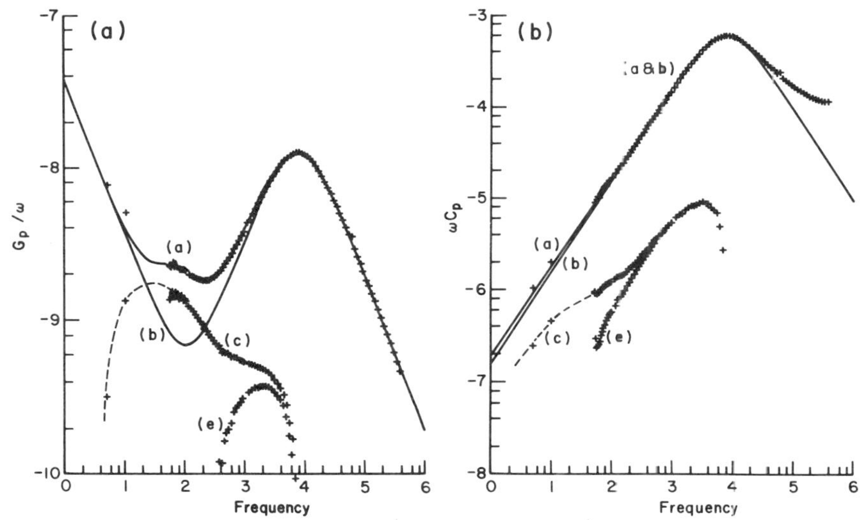

FIGURE 16. Experimental (crosses) and predicted (solid lines) G_p/ω *vs.* ω (Figure 16a) and ωC_p *vs.* ω response (Figure 16b) for the *n*-GaAs/$AlCl_3$–BPC interface. The bias voltage was 0.40 V and the $AlCl_3$–BPC mol ratio was ~0.8 : 1. Other labels and notation same as in Figure 6.

TABLE II. Parameters for the *n*-GaAs/$AlCl_3$-BPC[a] Mol Ratio ~0.8 : 1 Interface Determined by Automated Admittance Spectroscopy

	Parameter								
	C_{sc}, pF	G_{el}, mho	G_l, mho	C_e^1, pF	G_e^1, mho	C_e^2, pF	G_e^2, mho	N_e^1, $cm^{-2}eV^{-1}$	N_e^2, $cm^{-2}eV^{-1}$
Value	2.91×10^{-8}	1.23×10^{-3}	1.82×10^{-7}	3.70×10^{-9}	1.32×10^{-6}	1.45×10^{-9}	1.51×10^{-5}	1.21×10^{11}	4.8×10^{10}

[a] Values refer to rest potential of the system: 0.03 V *vs*. Al wire in 2 : 1 $AlCl_3$–BPC melt.

charge accumulation modes with individual R-C time constants are considered in parallel with C_{sc}.

A typical set of parameters for the n-GaAs/$AlCl_3$–BPC interface, determined from data such as those shown in Figure 16, using procedures outlined in Section 5, is assembled in Table II. These parameters (with the exception of G_{el}) are obviously voltage-dependent; those in Table II correspond to the rest potential of the interface (~0.03 V).

An estimate of the area density of states, N_e, at the interface responsible for charge accumulation is possible from the measured C_e values via Equation 30:

charge accumulation modes with individual R-C time constants are considered in parallel with C_{sc}.

Values of N_e thus determined for the two charge accumulation modes are also assembled in Table II. These values are in the range 10^{10}–10^{11} $cm^{-2}V^{-1}$ and are comparable to values reported by previous authors[27] for the n-GaAs/ aqueous electrolyte interface. This in turn supports the contention that such states are intrinsic to the n-GaAs surface and as such are not sensitive to the nature of the electrolyte contacting the n-GaAs surface. This conclusion from circuit modelling alone, however, must not be carried too far, particularly in the absence of supporting evidence from other measurement probes which are more sensitive on a molecular level. In many cases, a semiconductor surface may be altered appreciably in the presence of an electrolyte contact as, for example, when species from the electrolyte are specifically adsorbed on the semiconductor surface.

It was stated in Section 7.1 that interpretation of Mott–Schottky plots is frequently clouded by frequency dispersion effects. While a discussion of the causal effects for such dispersive phenomena is beyond the scope of this chapter, the dangers inherent in the interpretation of Mott–Schottky data obtained from measurements at *discrete* frequencies (which are often arbitrarily chosen) are illustrated by data on the test interface (Figure 17a) and model circuits (Figure 17b). These data were obtained from the *same experiments* as before (cf. Figures 14 and 2) but analyzed in a different manner. For these analyses, the parallel C_p value measured at each individual frequency was converted to its series equivalent, C_s. The Mott–Schottky data for three such frequencies are shown in Figure 15. Note that the assignment of unique *frequency-independent* V_{fb} and N_D values now becomes ambiguous *for the same system for which such values were defined by our modified analysis procedure*.

The point to note in Figure 17 is that the experimental data (crosses) coincide with the *frequency-independent* C_{sc} value (shown as dashed lines in Figure 17a and 17b) only over a narrow frequency range. Only when charge accumulation modes are completely absent in the system (exemplified by the response of model circuits a and b in Figure 17b) does the series capacitance remain constant with

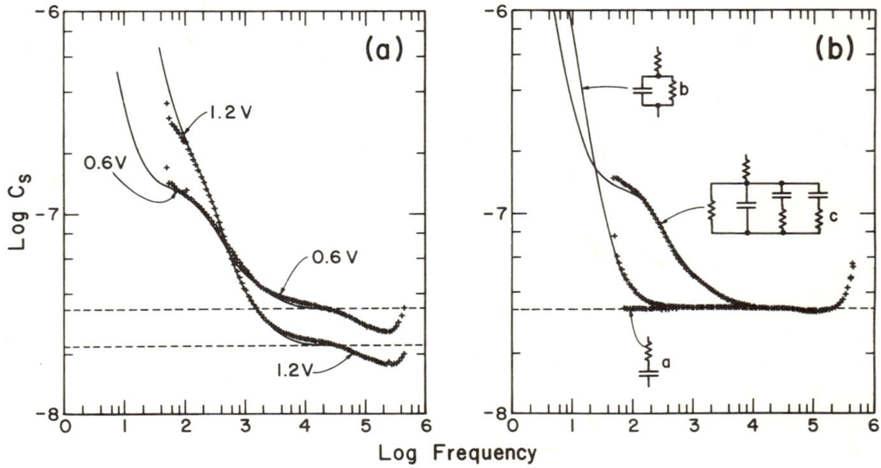

FIGURE 17. Experimental (crosses) and predicted (solid lines) C_s *vs.* ω response of the *n*-GaAs/$AlCl_3$–BPC interface at two bias voltages, 0.6 V and 1.2 V (Figure 17a) and of model circuits (Figure 17b). Parameters for generating the circuit response in Figures 17a and 17b were the same as those in Figures 14 and 2, respectively.

frequency. The departure of C_s from the C_{sc} value at high frequencies for the n-GaAs/$AlCl_3$–BPC interface (an effect *not* observed for the model circuits, cf. Figure 17b) may be caused by "fast" interface states not taken into account in the circuit models (note that this departure also shows up in ωC_p *vs.* ω plots for the interface, cf. Figure 16b).

The subsequent anomalous *increase* in the *G* values at very high frequencies (an effect seen both in Figure 17a and in Figure 17b) is due to amplifier "roll-off" which places an upper limit for the frequency range accessible for measurement (~0.1 MHz in the present case).

9.2. *The n-InP/$AlCl_3$–BPC Interface*

The importance of pretreating the electrode surface to remove flaws, surface damage induced by polishing, and other anomalies in Mott–Schottky measurements is clearly illustrated by a typical set of data in Figure 18 for the n-InP/$AlCl_3$–BPC interface.[26] This figure also presents a comparison of Mott–Schottky plots obtained from series capacitance measurements (using a lock-in amplifier, cf. Reference 28) at discrete frequencies (cf. Figure 17) with that based on frequency-independent C_{sc} values obtained by automated admittance spectroscopy. Note that while determination of V_{fb} is facilitated after electrode etching even for the C_s case, unique assignment of N_D is not possible. Unambiguous determination of *both* V_{fb} and N_D, however, is possible using the methodology presented in this chapter.

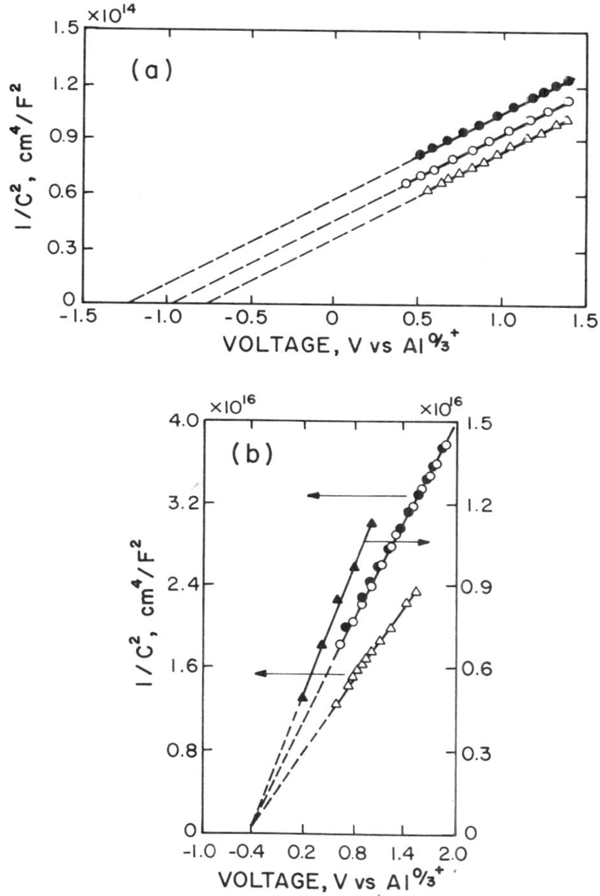

FIGURE 18. Mott-Schottky plots for *n*-InP electrodes (⟨111B⟩ orientation) in $AlCl_3$–BPC (0.85 : 1) electrolytes. The data in (a) and (b) are on electrode surfaces with no pretreatment and on etched electrodes, respectively. The ac signal frequency was 10 kHz, 6 Hz, and 1 kHz for (●), (○), and (△), respectively. Data labelled (▲) were obtained by automated admittance spectroscopy (copyright permission: Electrochemical Society, 1982).

Figure 19 shows G_p/ω *vs*. ω plots for the n-InP/$AlCl_3$–BPC interface. These plots are analogous to those presented previously for the n-GaAs case (cf. Figure 14). Note that similar to the GaAs case, the behavior of this system is completely described only by including two charge accumulation modes in parallel with C_{sc}. It is tempting to hypothesize that such an equivalent-circuit representation (cf. Figure 5a) is perhaps applicable to *all* III-V semiconductors in contact with an electrolyte. This representation also seems to be consistent with the unified defect model proposed by previous authors[29] for these materials. Further experimental verification of this hypothesis, however, is undoubtedly necessary. This is be-

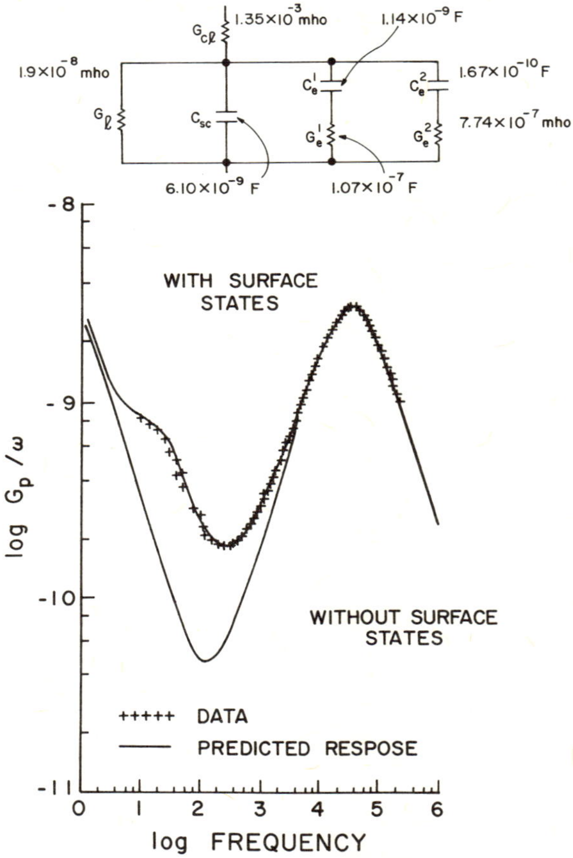

FIGURE 19. Plots of G_p/ω *vs.* ω for the *n*-InP/$AlCl_3$–BPC interface. The bias voltage was 0 V. The inset shows the equivalent circuit derived from these data (cf. Figure 5a). The $AlCl_3$–BPC composition was ~0.8 : 1 and the electrode orientation was ⟨111B⟩.

cause the situation when a semiconductor contacts an electrolyte is also considerably more complex than when the corresponding surface is in a vacuum.

10. Summary and Concluding Remarks

An automated technique has been described for admittance measurements of semiconductor/molten salt electrolyte interfaces. Equivalent circuit models have been developed to facilitate extraction of passive circuit elements from the measured equivalent parallel conductance and capacitance. Treatment of data in the form of G_p/ω *vs.* ω and ωC_p *vs.* ω plots, successfully employed by previous authors for the characterization of surface states at the silicon/oxide interface,[11] are shown herein to be equally convenient for the study of semiconductor/electrolyte interfaces.

Experimental results presented in this chapter for model systems (n-GaAs/$AlCl_3$–BPC and n-InP/$AlCl_3$–BPC interfaces) admittedly provide only a glimpse of the possibilities with this technique. Further clever experiments building upon the methodology developed thus far, for example, could explore correlations between equivalent circuits and surface phenomena (e.g., specific adsorption). Elucidation of the mechanistic aspects of electrode surface treatment and their influence on solar conversion efficiency is amenable to study by the procedures detailed in this chapter. Unanswered questions on these aspects remain in this technologically important area.[30]

ACKNOWLEDGEMENTS

The work described herein was the collective effort of a group of researchers under the author's direction among whom the names of P. Smith, T. Mraz, and R. Thapar deserve special mention. It is a pleasure to acknowledge numerous discussions and helpful suggestions, particularly in the initial phases of this research, with Professors Robert J. Gale of Louisiana State University and Joel B. DuBow of Boston University. This research was partially supported by a grant from the Solar Energy Research Institute, Golden, Colorado.

References

1. M. Sluyters-Rehbach and J.H. Sluyters, in *Electroanalytical Chemistry—A Series of Advances*, Vol. 4, A.J. Bard, Ed., Marcel Dekker, New York (1970), p. 3.
2. R. DeLevie and L. Pospisil, *J. Electroanal. Chem.* **22,** 277 (1969); see also references therein.
3. D.D. MacDonald and M.C.H. McKubré, in *Modern Aspects of Electrochemistry*, Vol. 14, J.O'M. Bockris, B.E. Conway, and R.E. White, Eds., Plenum Press, New York (1982), p. 61.
4. D.D. Macdonald, *Transient Techniques in Electrochemistry*, Plenum Press, New York (1977).
5. R.D. Armstrong, M.F. Bell, and A.A. Metcalfe, *J. Electroanal. Chem.* **77,** 287 (1977).
6. (a) V.A. Myamlin and Yu.V. Pleskov, *Electrochemistry of Semiconductors*, Plenum Press, New York (1967), Chapter 3, p. 279; (b) P.J. Boddy, *J. Electroanal. Chem.* **10,** 190 (1965); (c) H. Gerischer, in *Physical Chemistry—An Advanced Treatise*, Vol. 9A, H. Eyring, Ed., Academic Press, New York (1970), Chapter 5.
7. (a) A.D. Graves, G.J. Hills, and D. Inman, in *Advances in Electrochemistry and Electrochemical Engineering*, Vol. 4, P. Delahay and C.W. Tobias, Eds., Interscience, New York (1966), pp. 157–162; (b) D. Inman, A.D. Graves, and R.S. Sethi, *Electrochemistry*, Vol. 1, Specialist Periodical Reports, The Chemical Society, London (1970), pp. 215–222; (c) D. Inman, A.D. Graves, and A.A. Nobile, *Electrochemistry*, Specialist Periodical Reports, The Chemical Society, London (1972), pp. 109–116; (d) D. Inman, J.E. Bowling, D.G. Lovering, and S.H. White, *Electrochemistry*, Specialist Periodical Reports, The Chemical Society, London (1974), pp. 158–160; (e) A.D. Graves, *J. Electroanal. Chem.* **25,** 349 (1970); (f) A.D. Graves and D.

Inman, *J. Electroanal. Chem.* **25,** 357 (1970); and (g) R.J. Gale and R.A. Osteryoung, *Electrochim. Acta* **25,** 1527 (1980).
8. (a) Yu. V. Pleskov, in *Comprehensive Treative of Electrochemistry,* Vol. 1, J.O'M. Bockris, B.E. Conway, and E. Yeager, Eds., Plenum Press (1980), Chapter 6, p. 291; (b) I. Uchida, H. Urushibata, and S. Toshima, *J. Electroanal. Chem.* **96,** 45 (1979); (c) I. Uchida, H. Urushibata, H. Akahoshi, and S. Toshima, *J. Electrochem. Soc.* **127,** 995 (1980); and (d) R.J. Gale, A.J. Nozik, J. Cooper, and R.A. Osteryoung, *Coll. Czech. Chem. Commun.* **47,** 1794 (1982).
9. J.F. McCann and S.P.S. Badwal, *J. Electrochem. Soc.* **129,** 551 (1982).
10. S. Roy Morrison, *The Chemical Physics of Surfaces,* Plenum Press, New York (1977).
11. E.H. Nicollian and A. Goetzberger, *Bell. Syst. Tech. J.* **26,** 1055 (1967).
12. J.E.B. Randles, *Disc. Faraday Soc.* **1,** 11 (1947).
13. K. Rajeshwar, P. Singh, and J. DuBow, *Electrochim. Acta* **23,** 1117 (1978).
14. S. Kar, K. Rajeshwar, P. Singh, and J. DuBow, *Sol. Energy* **23,** 129 (1979).
15. R. DeGryse, W.P. Gomes, F. Cadon, and J. Vennik, *J. Electrochem. Soc.* **122,** 711 (1975).
16. F. Cardon and W.P. Gomes, *J. Phys. D.* **11,** L63 (1978).
17. E.C. Dutoit, R.L. Van Meirhaeghe, F. Cardon, and W.P. Gomes, *Ber. Bunsenges Phys. Chem.* **79,** 1206 (1975).
18. W.W. Laflere, R.L. Van Meirhaeghe, F. Cardon, and W.P. Gomes, *Surf. Sci.* **59,** 401 (1976).
19. M.J. Madou, F. Cardon, and W.P. Gomes, *J. Electrochem. Soc.* **124,** 1625 (1977).
20. W.H. Laflere, R.L. Van Meirhaeghe, F. Cardon, and W.P. Gomes, *Surf. Sci.* **74,** 125 (1978).
21. H.J. Hovel, *Semiconductors and Semimetals—A Treatise,* Vol. 11, R.K. Willardson and A.C. Beer, Eds., Academic Press, New York (1975), p. 58.
22. P. Singh, K. Rajeshwar, R. Singh, and J. DuBow, *J. Electrochem. Soc.* **128,** 1396 (1981).
23. K. Rajeshwar, *J. Electrochem. Soc.* **129,** 1003 (1982).
24. M. Freeman, M.S. Thesis, Colorado State University (1979).
25. P. Smith, M.S. Thesis, Colorado State University (1980).
26. R. Thapar, J. DuBow, and K. Rajeshwar, *J. Electrochem. Soc.* **129,** 2009 (1982).
27. P. Janietz, R. Weiche, J. Westfahl, R. Landsberg, and R. Dehmlow, *J. Electroanal. Chem.* **106,** 23 (1980).
28. P. Singh, K. Rajeshwar, J. DuBow, and R. Job, *J. Am. Chem. Soc.* **102,** 4676 (1980).
29. W.E. Spicer, P.W. Chye, P.R. Skeath, C.Y. Su, and I. Lindau, *J. Vac. Sci. Technol.* **16,** 1422 (1979).
30. R. Noufi, private communication (1982).

Index